Who goes Canoeing with their Mother-in-Law?

THE MISGUIDED TALES OF AN AVID PADDLER

Written by KYLE PENNER
Illustrated by PATTI ENNS

◆ FriesenPress

Suite 300 - 990 Fort St
Victoria, BC, V8V 3K2
Canada

www.friesenpress.com

Copyright © 2019 by Kyle Penner
First Edition — 2019

Illustrated by Patti Enns
Author Picture by David Klassen

All rights reserved.

Disclaimer: Warning: Paddle at your own risk. Always wear a PFD. This book is not intended to keep you safe. The maps provided are an artist's rendition of the routes. They are not to scale, skip some details, and should not be used as backcountry guides. Please read chapters 1, 4, and 6 to figure out why you should always use a proper topographical map.

For speaking arrangements, bulk sales, and more info,
please visit www.kylepenner.net.

No part of this publication may be reproduced in any form, or by any means, electronic or mechanical, including photocopying, recording, or any information browsing, storage, or retrieval system, without permission in writing from FriesenPress.

ISBN
978-1-5255-5737-8 (Hardcover)
978-1-5255-5738-5 (Paperback)
978-1-5255-5739-2 (eBook)

1. BIOGRAPHY & AUTOBIOGRAPHY, ADVENTURERS & EXPLORERS

Distributed to the trade by The Ingram Book Company

Endorsements

"Who Goes Canoeing With Their Mother-in-Law? has all the charm and adventure of a week long trek into the Canadian woods. Kyle Penner is the best of travelling companions: relaxed, keen -- and amusing."

- Roy MacGregor,
author of Canoe Country: The Making of Canada

"Just getting into a canoe can be an adventure in itself, but paddling off into the wilderness with your mother-in-law pretty much ups that likelihood to a certainty. I hope you enjoy reading Kyle's canoeing tales as much as I did!"

- Dana Starkell, *Paddle to the Amazon*

"Kyle Penner's stories read like a downhill portage after the peak of blackfly season. They're a joy to devour, and exhilarating enough to want to pore over again and again."

- Kevin Callan *(a.k.a. The Happy Camper),*
author of Once Around Algonquin: An Epic Canoe Journey.

"The stories have a familiar feel that every outdoorsman can relate to but with a twist that kept me wanting to read more! Definitely allows you to slip into the wilderness from the comforts and security of home! What a great read!"

- Mike Ranta, *author of Mike and Spitzii's Great Canadian Adventure: Cross-Continental Coast to Coast Record-Breaking Solo Canoe Expedition*

Entertaining and singularly wild.

— Peter Keleghan,
actor, the Red Green Show

Kyle Penner tells a good story with a light-hearted comic charm. Here we have stories on many a common paddlers theme and with much learning gently shared in such a relaxing way that it is easily absorbed — from how to wisely canoe camp with young children to dealing with those tense wind bound moments. There is universal appeal to these stories but they will particularly resonate with local paddlers in Southern Manitoba. Beautifully illustrated, this avid paddler will inspire you too, as Henry David Thoreau once put it, 'Everyone must believe in something. I believe I'll go canoeing.'"

— Bob Henderson, *canoe tripper, educator, author of Every Trail Has a Story, www.bobhenderson.ca*

Immediately I noticed the light prose that gave life to every adventure, and adventures there are from mothers-in-law to Bigfoot. I have paddled these trips but now I have a new view and appreciation of the routes and challenges. I recognize the landscape, and myself, in many stories. These are delightful tales of trips made by Kyle providing insight into paddling, trips, and misadventures (to be avoided). The beautiful artwork of routes draws the reader into each story. This is a must have for paddlers in Manitoba.

— Charles Burchill, *Analyst, Biologist, Canoeist*

Kyle has a warm, friendly approach to writing about his canoeing adventures. I felt like a close friend looking forward to joining Kyle on his next trip so that we could swap more stories. He can turn an ordinary two-day canoe trip into a memorable event with his characterization of people and detailed descriptions of places. Patti's drawings provide quaint, visual context to the locations that Kyle writes about. I joyfully anticipate reading more of his stories.

- Mel Baughman, *forester, wilderness canoeist, adventurer.*

This is a fun read. Kyle's stories are homey, engrossing, informative, and memorable. Just the thing to keep experienced paddlers interested, and encourage new paddlers to search out their own wilderness adventure. A great addition to anyone's outdoor library.

- Doug McKown, *author of Canoeing: Safety and Rescue*

I've sometimes been asked: 'What is a normal day on a canoe trip?' My response usually goes something like: 'There really are no normal days on a canoe trip'. Kyle certainly shows that to be true. He shows that the normal day on a canoe trip is filled with things that one likely did not anticipate. That is what makes canoe tripping magical. There is this unplanned magic in Kyle's stories.

- Ric Driedeiger, *Churchill River Canoe Outfitters*

Table of Contents

Introduction	1
1 — Don't Yell At Meteorologists	3
2 — The Grudge	15
3 — The Jolly Canuck Riparian Catering Service	25
4 — The Lost Hikers	31
5 — Stuck Between a Rock and a Hard Place	47
6 — Who Goes Canoeing with Their Mother-in-Law?	57
7 — Bigfoot is Real	67
8 — But I Saw a Waggle!	81
9 — Somebody Stole Our Canoe	97
10 — The Advantage of Working on Sundays	117
11 — Paddle Faster, I Hear Banjoes!	125
12 — Taking Good Notes	133
13 — Aliens are Coming to Abduct Us	143
14 — Me and My Boys	155
15 — Seven People in a Canoe	161
16 — Windbound	169
17 — What Do You Mean You Don't Like Portaging?	187
Acknowledgements	**205**

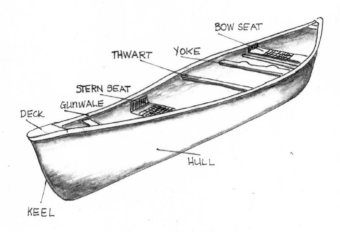

I acknowledge that I have paddled extensively through Treaty Territories 1, 2, 3, and 5, the traditional territory of the Anishinaabeg, Cree, Oji-Cree, Dakota, and Dene Peoples, and on the homeland of the Métis Nation. For the gift of sharing this beautiful land, I am thankful.

To Ashley,
If I had to pick the perfect paddling partner,
it wouldn't be your mother.
It would be you.

Introduction

July 2018

The six of us dipped our paddles into the Rice River at an easy pace, the bald eagles watching us from their tree thrones high above. When we paddled too close, we'd hear their wings beat against the air as they took flight, upset that we would dare disturb their watch.

I heard a soft splash as Nick's canoe approached from behind. As he neared, I found myself recounting the many adventures I'd had canoeing. I could not stop talking. There were times I got lost, plus that trip when my canoe floated away. There was being rescued, that one trip I canoed with my mother-in-law, and that other time I had forgotten my map. Let alone the Bigfoot stories.

Nick graciously listened to my rambling—until an epiphany struck me. I stopped talking and took a breath.

"You know what, Nick? I really need to write all these stories down."

"Wait, what?" Nick looked bewildered. "You went canoeing with your mother-in-law? Who does that?"

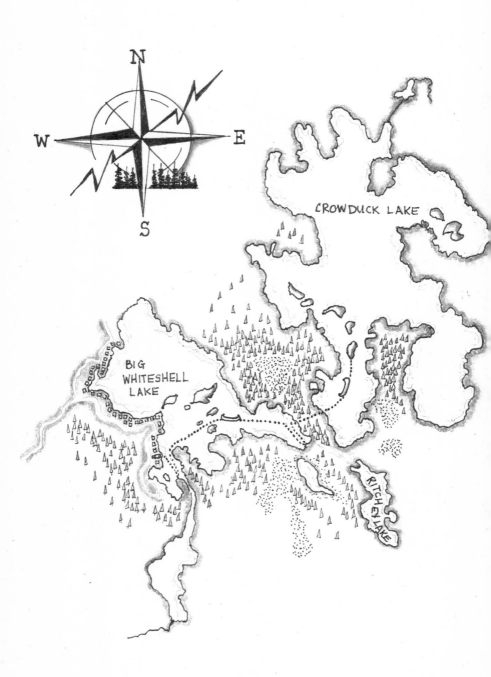

− 1 −

Don't Yell At Meteorologists

July 2004

When you decide to go on a canoe trip, don't stay up until 5:00 a.m. the night before you leave.

But alas, that's what we did.

My paddling partner was David. He was my roommate, bandmate, camp co-worker, locker partner, and friend throughout high school.

The year was 2004, back when you had to connect your video game consoles to each other with cables instead of playing online. A considerable amount of work went into getting a house full of people together to play on four separate gaming systems. And, because of all that hard work, we figured we needed to make the most of it. So we played video games until 5:00 a.m.

Needless to say, our plans to hit the road bright and early didn't quite pan out. We left Winnipeg at the crack of 1:00 p.m. and made it to Rennie, just outside the Whiteshell Provincial Park, a sluggish hour and a half later. Ashley and I were only dating at the time, and Rennie was where her parents lived.

We had planned to borrow their canoe. If you could even call it a canoe. It was an old 14-foot clunker they had bought off the internet, complete with a fiberglass body, metal thwarts and gunwales, plywood for seats, and just a hint of its original red paint. And it was heavy. Like, 60 kilograms heavy. This boat was a beast, a battleship, designed for floating and little else. There are places in this world for old boats that fill our hearts with warm nostalgia, but this was not one of those boats.

But since we were both poor university students at the time, what else could we do? We strapped the red clunker to the top of my 1994 Ford Tempo nicknamed "Clara the Silver Bullet" and headed to Big Whiteshell Lake.

Two years earlier, David and I had worked at summer camp together. On a weekend off, our directors organized a "guys fishing weekend" at Crowduck Lake. David and I didn't know much about Crowduck Lake other than there was supposed to be good fishing there. We helped load up a bunch of fishing boats, along with motors, tents, food, rods, water jugs, tarps, backpacks, and fishing gear onto the back of a trailer, and off we all went.

What our directors had failed to tell us was that you can't actually drive to Crowduck Lake. You can drive to Big Whiteshell Lake, but then you have to boat across the lake, find a little triangle marking the portage, and carry your gear 750 metres into Crowduck Lake. The first half of the portage was an uphill haul that our prairie legs were not prepared for. The second half was a muddy slog that had already claimed its fair share of shoes from previous paddlers.

What our directors also failed to tell us was that, because we were young, we would be the ones tasked with lugging the aluminum fishing boats and motors on that harrowing journey. When planning the trip, I'm sure they were wondering how to get all that gear into Crowduck Lake, until one of them looked up from the map and said, with a gleam in his eye, "Wait a minute! We have a bunch of 19- and 20-year-olds who can carry all the heavy stuff!" An evil laugh would have followed as they rushed out the door to invite "the guys" on a fishing trip.

Now, two years later, David and I were that much older and that much wiser. Wanting nothing to do with hauling aluminum fishing boats, we headed out to Crowduck Lake in a canoe, thus making the portage much more bearable. Unfortunately, the canoe we had was the old red clunker, so despite our age and wisdom, we weren't actually carrying much less weight than on the previous trip.

We figured we'd start the trip off right with a good meal, so we stayed in Rennie for supper with Ashley's parents. It was the July long weekend, and we knew we'd have sunlight until at least 10:00 p.m.

By the time we got our canoe loaded and into the waters of Big Whiteshell Lake, it was 7:00 p.m. We had three hours to paddle five kilometres across Big Whiteshell Lake, portage our gear, and then paddle a few more kilometres to an island to set up camp. What could go wrong?

For starters, we didn't have a map. Having one of those might have been helpful. We were depending entirely upon our memory, which is a terrible idea when you were previously just a passenger in a boat.

With youthful confidence I steered our boat south, giving no thought to the easterly direction we should have taken. We had travelled over two kilometres before we realized we were heading the wrong way. But no worries! We still had lots of daylight!

We paddled north back to our starting point, and from there headed straight east like we should have in the first place. The lake was quiet, and the sky was fading to orange as the sun slowly set behind us. The wind had died down, the motor boats had packed it in for the night, and the bugs hadn't found us yet. It was just us out there in the middle of the lake. Us, and the soft sounds of our paddles dipping in and out of the water.

With our spirits high, our voices cut through the still evening air.

"Hey, David. Remember when the weather report said 'scattered thundershowers'? Ha! Look how calm and peaceful it is out here."

"Stupid weather experts! They never know what they're talking about."

A few more quiet strokes, and our youthful bravado was miles high.

"Yeah, stupid meteorologists!"

"Screw you and your wrong weather reports!"

"What do they even pay you for?"

CRACK!

Our paddles froze and our heads whipped around.

"Holy crap, Kyle, we need to paddle. Fast!"

Creeping up to cover the sun behind us was a wall of black clouds heading straight for us. A streak of lightning blasted the mainland behind us.

CRACK!

We were only a few kilometres ahead of the storm, in the middle of the lake, with two kilometres to go to our portage point.

We dropped to our knees and dug our paddles in as fast as we could.

The wind was picking up, and the thunder cracking behind us drove us to paddle at an even more frenzied pace.

CRACK!

"Paddle harder, David! Keep going!"

Every few strokes we'd turn our heads to steal glances at the oncoming storm. The wall of water had reached the western lakeshore and was advancing to wash away what was left of our confidence.

CRACK!

"PADDLE!"

Our breaths had shortened to pants when the rain hit us. Turning our heads, we couldn't see the opposite shore. It wouldn't be light for long.

The eastern shore loomed ahead of us. We'd be close to getting off the water right when the lightning would surround us. As we neared the tree line, a bright light shone out at us from behind the trees. It flashed on and off repeatedly.

"Kyle! That has to be a flashlight! Someone's signaling us!"

Behind the flashing light, and through the leaves on the shoreline, the flames of a campfire flickered.

"But there aren't supposed to be any campsites here!" I shouted, struggling to make myself heard over the wind and the rain. "Besides, this is the wrong landing! We want the portage to Crowduck Lake, but this is the portage to Ritchey Lake!"

"So what do we do?"

We glanced behind us at the oncoming lightning.

"We head to the correct portage. It's not that far. We'll set up our tent there, under a tree."

"What about these campers?"

"Leave them! We can make it."

CRACK!

The sky was growing darker. Our paddling was now frantic. The wind howled.

After a kilometre we saw the fluorescent wooden triangle made of two by fours, marking the portage.

"There, David! I see it! We made it!"

Our muscles relaxed. Our breathing lessened.

As we drew near the portage, a flash of lightning illuminated the lettering on the bottom of the triangle.

Ritchey Lake: 2 km.

We were at the wrong portage.

"You've got to be kidding me! Those other campers must be at the Crowduck portage!"

"Maybe we should have brought a map?"

"Shut up, David!"

CRACK!

"What are we supposed to do now?"

"Let's go back to those campers. They had a fire. At least we'll be warm."

Turning the boat around, we knew we were now heading straight into the fierce wind, sheets of rain, and frequent flashes of lightning.

We clenched our jaws and readied our wills.

"Let's go!"

With every plunge of our paddles we yelled "Stroke!" We set a rhythm that would have made the voyageurs proud. We leaned forward for each pull and dragged the water back with all our strength. We were flying. Well, relatively speaking. We were still in a 14-foot clunker of a canoe, after all.

The lightning bolts had surrounded us now with almost constant CRACKS! Darkness enveloped us.

We pressed on, rounding a jut of land, and saw the campfire again. It was barely clinging to life amidst the rain, but still flickering. Warmth and rest were close.

The rain was now a waterfall, crashing down on our heads. I could hardly see David sitting two metres in front of me. With no time to don our rain gear, we were drenched from head to toe.

"A few more strokes David!"

We felt our canoe scrape against the lake bottom, bringing our boat to a halt. Jumping into the shallow water, we pulled the canoe a few feet up the shore and embraced under the nearest tree.

We couldn't help it. We started jumping around, holding each other's arms as we celebrated. We made it! We were alive!

CRACK!

"Right. Let's go see what the deal is with those campers."

Finding a path to the campsite, we called as we came up to it. "Hello? Is anybody here?"

"Bonjour!" the reply came from inside the tent, followed by a string of French that neither of us could understand.

"Oh. Je suis desolé. I'm sorry. Our French is quite terrible. Do you by chance speak English? Parlez-vous anglais?"

"I said, what the hell are you guys doing out there in that storm? We tried signaling you to come here!"

"Uh, yeah. We must not have seen it."

We looked around the campsite. There was one tent, lit up from the inside with a lamp. They had pulled up their fishing boat through the shrubs alongside the shore. An orange tarp was strung up between some trees, and under it were their camping chairs, coolers, kitchen supplies, firewood, and a massive pile of crushed beer cans.

David and I looked at each other and shrugged.

"We're a bit wet, and it's late. Can we camp here with you for the night?"

"Oui, oui! Of course! We'll come out when the rain dies down. Throw some logs on the fire to keep it going!"

We stoked the fire and walked back to our canoe to grab our bags. The rain was starting to thin out. David set up the tent, which had somehow stayed dry, while I sat under the orange tarp digging through our bags for dry clothes.

The storm must have been moving fast, because the rain ceased as soon as we had climbed into our tent and changed out of our soaked clothing.

We were warming ourselves by our neighbour's fire when the tent door beside us unzipped. Out came a short, older, bearded man wearing a flannel shirt and suspenders. Behind him was a taller man, clean shaven with darker hair, wearing a white t-shirt and jeans. Last but not least came a boy about twelve years old, his eyes darting between the two men he knew, and the two men who were intruding upon his campsite.

The first man introduced himself to us, shaking our hands. "Hi, I'm Jean." He pointed to his friend and the boy. "And this is Luke and his son, Isaac."

"I'm Kyle, and this is my friend, David." David and I stood by the fire warming our hands, and Jean, Luke, and Isaac sat down.

"Do you guys want a beer?" Jean asked, gesturing toward the cooler.

We declined, knowing alcohol on camping trips comes in limited quantities. We didn't want to loot their treasure.

Luke glanced over at Isaac. "Boy, get me a beer."

Isaac scampered over, opened two cans of beer, and handed them over to the men. Luke chugged his and threw the can into the now-growing pile of empties. Jean sipped his at a more socially acceptable rate.

"There, that's better." His eyes rolled back as he lit a cigarette.

David and I looked at each other. They were definitely drunk. And where they were camping was not legal. It appeared to us that they had just picked a spot on the lake and started camping. Jean threw a few cut logs into the fire. Watching them burn, I noticed their perfectly sawed off ends. These logs were made by a chainsaw! I peered around the campsite and quickly concluded that they had cleared a random spot in the forest and set up camp.

But, without their illegal campsite and their warm fire, we'd probably be setting up our tent in a puddle under a tree somewhere. So we shrugged and embraced it.

The next morning we said our thanks to them and were on our way. Well, actually, we said our thanks only to Jean, as neither Luke nor his son Isaac emerged from their tent that morning.

As we gathered our gear back at the portage, David and I shook our heads at our incredible luck that these buffoons were there that night, and how their buffoonery had saved us from a night of misery.

With full stomachs, warm clothes, and blue skies above us, we started portaging into Crowduck Lake. It was just as we remembered. Uphill at the beginning. Muddy at the end. And two of us carrying that red canoe was no better than four of us carrying an aluminum fishing boat.

But we made it to Crowduck Lake! We found an island, set up our tent, threw our fishing lines into the water, and caught some pickerel. Compared to the previous day it was blissfully uneventful. We went to bed rather early that night, still tired from the night before.

The next day, we started back. When we finished the muddy portage, we looked to our right where Jean, Luke, and Issac had been camping. Other than a fire ring with half burned logs in it, the bent shrubs along the shore from the boat being dragged over them, and a whole bunch of trees bearing chain saw cuts, there was no evidence of them being there. Okay, so that was a lot of evidence. Clearly, they weren't "leave no trace" campers. At least they had spent a few minutes cleaning up their pile of beer cans. That counts for something, right?

We paddled straight west, riding a tailwind that made paddling a breeze. In the car, we debriefed our canoe trip.

"Well, that didn't go quite as expected, did it?"

"No. No it did not."

"Next time, Kyle? Just bring a map."

− 2 −

The Grudge

July 2004

The Apostle Paul once said, "Bear with each other and forgive one another if any of you has a grievance against someone."

It took a canoe trip, and half a decade, to learn what this actually means.

In the summer of 2004 I was completing my pastoral internship at Grace Mennonite Church. I preached a few sermons, went to a few meetings, visited some seniors in nursing homes, and planned a summer trip for the youth.

Of course, the trip I planned was a canoe trip.

Our large group of sixteen included five adults and eleven teenagers. The adult leaders who had volunteered all year with the kids were Phil, James, Ashley, and myself, and we were all keen for this canoe trip. For safety reasons we wanted a nurse and certified lifeguard to come along, and, since our friend Patrick was both of those, we brought him along too.

Due to the size of our group, plus having some rookie paddlers, we had to pick a fairly straightforward trip. We chose the Caddy Lake tunnels route. The maximum paddling time each day would be a couple of hours. There were big campsites along the way. We expected to spend one night next to a large waterfall. And the first day would involve paddling through some tunnels. It was perfect for a group of teenagers.

Over 100 years ago the Caddy Lake tunnels were blasted through solid granite rock beneath the two national railroad tracks, the CNR and the CPR, to help the flow of the Whiteshell River downstream. Nowadays, they are also a unique and popular paddling destination with numerous backcountry sites in close proximity.

It was a bright, clear day in July, and within an hour of setting out most of the kids had figured out how to keep a canoe straight. A few of them picked up the J-Stroke technique, and the rest learned to use the stern paddle as a rudder. All except my fifteen-year-old little brother Darcy. In the bow of his canoe was Kyle, a seventeen-year-old ox of a teenager. They seemed content to zig-zag their way up the lake, oblivious to the fact that they were working twice as hard as the rest of us.

With blue skies and little wind, the weather was sublime. Paddling through the tunnels is always a highlight, no matter how many times you've done it. Entering the first tunnel from Caddy Lake is akin to entering a small rainforest—ferns growing along the edges, the sun blocked by leafy towering trees, and birds flitting back and forth from their nests in the cave. The second tunnel contrasts the first—its length and narrowness make it more ominous. It flows faster, and the roof slopes in the middle, ready to scrape any unsuspecting heads. Because of its length, this tunnel is so dark that you can't see the bow of your canoe. The only light is from the cave entrance and exit.

WHO GOES CANOEING WITH THEIR MOTHER-IN-LAW?

We camped on a little island on North Cross Lake the first night, a four-hour paddle from our launch point. The island was dotted with the bright colours of all our tents, the blue coolers stacked beside the picnic table, and eight red canoes flipped over on their sides, resting after a hard day's work.

Ashley had taken it upon herself to organize the food. Not only did she make a menu plan and figure out the quantities needed for sixteen people, but she had also sorted and labelled each cooler by meal. Arranging the food for canoe trips this way has proven to be not only a time saver, but also ensures the food stays cold by not having to open coolers all the time.

Everybody was put into a meal prep and cleaning group, and already on the first day camp ran like a lean, mean, well-oiled machine.

That night, we thought we must be close to a record for one of the coldest nights in July. The temperature reached low single digits, and throughout the night we could see our breath puffing out in front of us.

A hearty breakfast and some hard work packing up camp got the blood flowing, just in time for a day of paddling.

Much of it was the same as the day before. There were peaceful moments when we could hear the birds sing, and boisterous moments when the kids sang "Hey Jude" by The Beatles at the top of their lungs. We stopped often to let the last boat catch up to us, which was usually Darcy and Kyle zigzagging away. Every hour we'd pull all the boats together into a flotilla and crack open the appropriate coolers to share the snacks between the canoes. The kids discovered quickly that, if you are the person who picks out all the Smarties from the bag of trail mix, you are no longer trusted with any other important snacks.

While eating all the Smarties out of the trail mix bag is a pretty selfish move, it is worth noting that much of a canoe trip is learning how to work with the different personalities that have come along. Some of us canoe to de-stress, but for others being on the water in the wild can be the stressful part. Throw in group dynamics, group decision-making, a couple of whiners, a few insecure folk, a couple overconfident ones, some strong paddlers, weak paddlers, and people who wonder why they're even paddling in the first place, and things can get volatile. Thankfully, cooler heads prevailed over the Smarties fiasco that day.

Our second night was spent at Mallard Falls, a long waterfall at the end of Mallard Lake. There aren't a lot of oak trees growing in the Whiteshell Provincial Park, yet the portage/campsite at Mallard Falls has oaks growing everywhere. It certainly is an anomaly. I've heard it said that wherever First Nations historically camped you'd find oak trees growing, as some of the acorns they didn't consume would have started to grow. That makes sense for Mallard Falls, as Mallard Lake is filled with wild rice and one would have to portage around the falls every year at harvest time.

After another flawless meal prep and cleanup organized by Ashley, we fell asleep that night to the gentle roar of the waterfall beside us, taking comfort that we weren't the first people to sleep under these oaks.

The next day ran smoothly again until it was time to pack up. Partway through our regimen of deconstructing tents, washing up morning dishes, and loading everything back into the canoes, Phil and James uncovered an environmental disaster. Throughout the previous day, our group had found a log to sit on to go to the bathroom but hadn't bothered burying their business. It was a heap of human ignorance, sure to irk the next groups to pass by. Phil and James, approaching sainthood, took our camp shovel, stoked the fire, and made the most out of a very crappy situation.

With only ten kilometres to the island campsite on Lone Island Lake, day three of paddling was relatively short. The weather was cooperating, the snacks were delicious, and Darcy and Kyle were still happy zigzagging their way around. In between Mallard Falls and Lone Island Lake you have to paddle the fairly narrow and winding Whiteshell River. For those of us who could paddle in a straight line, we could easily fit two or three canoes abreast. But Kyle and Darcy kept bumping into the reed-filled muskeg shore, and then into the other shore, and then back to the original shore.

Phil and I offered to teach them how to J-stroke or rudder to keep their canoe straight. We even offered to switch places so that somebody who could steer was in the back of their canoe. Darcy and Kyle stared back at us with the disdain only teenagers can muster and held up their water bottles.

"We're doing this on purpose! We're collecting snails!"

And, sure enough, each of their water bottles was filled with snails. We didn't ask if they were drinking that water. We figured they were both responsible teenagers, and any trauma they inflicted on their digestive system would be an opportune learning experience.

That night we set up camp on the lone island in Lone Island Lake. I was on the supper prep and cleanup crew. We enjoyed our hot dogs and vegetables, and near the end of our meal I looked at the menu plan for dessert.

S'mores—the delectable combination of a roasted marshmallow and a piece of Jersey Milk chocolate bar melting between two graham crackers. My stomach lifted in anticipation.

I opened cooler number seven to get the supplies.

I pulled out the bag of marshmallows. I dug around for the graham crackers and pulled the plastic packages out of the blue cardboard box. I reached down into the corner for the three Jersey Milk chocolate bars. When my fingers took hold of them, I realized there were only two and a half Jersey Milk chocolate bars in my hand.

"Ashley! Why is half a chocolate bar missing?"

"What do mean?" she asked, and walked over to search the cooler herself. "Oh, come on! Somebody ate it!"

I was livid.

We had packed just the right amount of food per meal, and Ashley had a system in place to ensure that we only ate certain food on certain days. There are unspoken rules on canoe trips, and not raiding the next day's supply of food is one of them. What an inconsiderate, rude, uncaring, terrible, awful, selfish thing to do.

I lost it.

"EVERYBODY!" I yelled. "SIT DOWN BY THE FIRE!" The kids sat down and looked up at me. The other adults stopped talking and stared, wondering what had roused my anger.

"Somebody ate this chocolate bar and I want to know who. Which one of you is so selfish that they would eat the next day's food? Tell me who did this!"

Silence.

"Which one of you thinks they're that much more important than the rest of us? Huh?"

Blank faces looked at me. Some eyes were downcast to avoid my manic glare.

"What is wrong with you? We had a system. Why couldn't you follow it? Speak up, and then the rest of you can go on your way."

Nobody moved. Except for Phil. He moved in close and whispered.

"Kyle. This isn't going to work. They're not going to admit it was them in front of everyone. You've backed them into a corner. Help them save face. Give them an out."

"Fine." I turned back to the group. "Alright. If you ate the chocolate bar, or if you know who did, you can come to me later and we'll work something out. But because of this one person, we have to ration the chocolates for everyone's S'mores. I hope you're happy."

I went to bed grumpy that night. Grumpy that the guilty kid didn't step forward, and grumpy that the marshmallow to chocolate ratio of my S'more was woefully inadequate.

The next morning, before breakfast, I opened the final cooler packed with our last breakfast of bacon and eggs, and let me tell you, that kid was lucky they didn't eat any of that meal too, because then I would have really lost it.

We paddled north on Lone Island Lake, and as the day went on the least outdoorsy kid was losing his patience with this whole "canoeing business." He was in my boat, and right near the start of the day he turned around and asked "How long are we paddling today?"

"Oh, I dunno. 'Bout a good hour."

Roughly thirty minutes later, we were still in the middle of the lake, paddling into a gentle headwind. He turned around and asked again.

"How much longer, Kyle?"

"Hmmm…About a good hour."

At the end of the lake, a few kilometres from the island, there's a boat launch. He got quite animated when he saw it and asked, "Is that it? Are we done?"

"No, sorry. James's dad is meeting us at the next picnic spot."

"Well, how far away is that?"

"Look," I said. "Nothing I say will get us any closer any faster. The only thing we can do is to keep paddling. We'll probably get there in about, oh, a good hour."

Frustrated silence emanated from the front of the canoe as we paddled past the boat launch and into the river.

Fifteen more minutes passed.

"How much longer, Kyle?"

"A good hour."

"NO!" he yelled, slamming his paddle down across the gunwales. "KYLE! THAT'S IT! I'M DONE!" He had turned to face me now, his eyes wide and his face turning red. "ALL I WANT TO KNOW IS HOW MUCH FURTHER WE HAVE 'TIL I CAN GET OUT OF THIS STUPID CANOE! IF YOU SAY 'A GOOD HOUR' ONE MORE TIME I'M GOING TO THROW MY PADDLE AT YOU! JUST TELL ME HOW MUCH FURTHER IT IS SO I KNOW!"

I smiled.

"Turn around, man. See that van there? With the canoe trailer? We're here."

I received no smiles for my sadistic game. Only grunts and glares.

Five years later Ashley and I were running a camp on Matheson Island, a fishing community just north of the Lake Winnipeg Narrows. We were at a different church now, with a different set of teenagers, but we needed a nurse and lifeguard for the week, so we asked our friend Patrick to come and help out again.

On the last night of camp we were having a campfire, and Ashley and I brought out a cooler with the supplies for S'mores. With all the teenagers gathered around the fire, I started lecturing them.

"Before I give you these S'mores, I have a story to tell. One time, I was canoeing with a different group of teenagers, and one of them selfishly ate some of the chocolate bars the day before we were supposed to have S'mores. And they never fessed up."

I heard snickering behind me. I whipped my head around to see Patrick covering his mouth, holding back laughter.

"What's so funny Patrick?"

"You're still mad about that?" He burst out laughing. "It was me!"

"You!?!? Why would you eat the chocolate bar?!"

"BECAUSE I WAS HUNGRY!"

"Why didn't you admit it was you?"

"Because you were so angry! I didn't want you to yell at me!"

"You're a terrible person, Patrick."

"For eating half a chocolate bar? I think maybe you're the terrible person for hanging on to this grudge for five years."

And you know what? He was right.

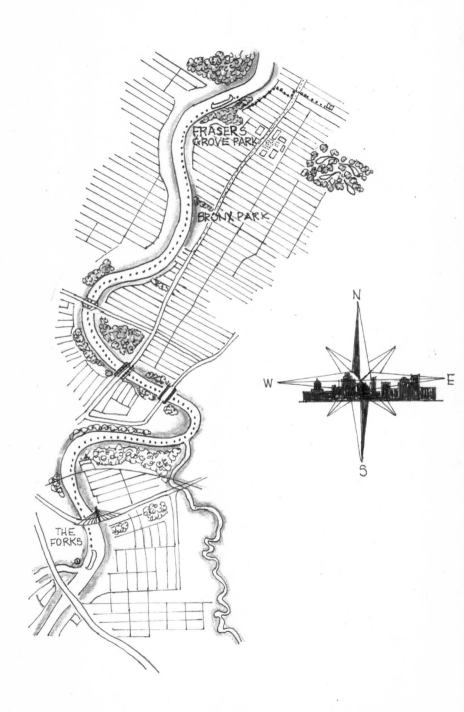

– 3 –

The Jolly Canuck Riparian Catering Service

July 2008

On Canada Day I had a dream come true. What made it even better was that, when I woke up that morning, I didn't even know this dream existed.

Ashley and I were living in Winnipeg at the time, and our friend Chris ended up spending the night at our house after an Ultimate Frisbee game.

In the morning, Chris and I decided to paddle the infamous Red River. We put on our life jackets, filled up some water bottles, attached paddles to my canoe using bungee cords, and started my first urban portage. The portage was one kilometer long, but Henderson Highway, the main traffic vein through northeast Winnipeg, was blocking our way. We stopped at a bus stop and lifted the canoe high above our heads to look for traffic. All six lanes were clear, so we lowered the canoe back over our heads and blindly scampered across the highway. After passing the local credit union and two blocks of housing, we were under the mighty elm trees of Fraser's Grove Park. We slid the canoe down the river bank into the Red River and squished our way through the mud.

We pointed the canoe south and were off.

Our destination? The Forks National Historic Site, where the Red and Assiniboine Rivers meet. The Forks has been a meeting point for thousands of years, from the First Nations to the fur traders to thousands of immigrants. And, this day, it would play host to Chris and me.

When you paddle a river, intending to return to your starting point, always start paddling upstream. Always. Thankfully, I've never had to learn why the hard way.

If you've never had a chance to paddle on a river through a city, I'd highly recommend it. Especially the Red River in Winnipeg. Because the water level fluctuates up to six metres a year due to spring flooding, most of the property along the shoreline is undeveloped. The trees are massive. Elms, cottonwoods, and Manitoba maples are everywhere, oblivious that they're growing in the middle of a thriving city. Between the scarcity of signs or familiar landmarks, and the miniature forests that make up most of the river bank, even pinpointing your exact location in the city is tough. We could only orient ourselves when we paddled under bridges.

Having never paddled on the Red River through the city before, we were taken aback by not only the dense greenery, but also the hidden beaches, the wildlife, and the number of shopping carts we saw poking up along the banks.

Along the way we passed Bronx Park. Overlooking the river is a plaque recognizing Don Starkell and his two sons, Dana and Jeff, and their paddle to the Amazon River. The Starkells started that famous trip from this park, and their book *Paddle to the Amazon* was one of the inspirations for me to pick up a paddle in the first place. It's such a wonderful quirk that the city of Winnipeg has a plaque commemorating this journey and the phenomenal stories that Don Starkell recorded, even if he was a tad unorthodox at times.

We paddled for two hours and arrived at the Forks. It was Canada Day, so the place was bustling with activity. The restaurant patios were full for lunch, the buskers were strumming their guitars, and some teenagers were making out on the grass.

We turned into the Assiniboine River and noticed several boats parked at the federally owned docks. As we pulled up to them, I shouted, "Chris! Is that a hot dog boat?"

Before Chris could answer, we were greeted by a voice from inside the boat. "You bet it is!"

What was this? Were our long and laborious lives being rewarded with a hot dog boat serving jumbo dogs to hungry canoeists? Had we died and gone to heaven?

While it wasn't quite heaven, it certainly was right up there in divine experiences. We had stumbled upon the Jolly Canuck Riparian Catering Service, a boat retrofitted into a glorified hot dog cart. Yes, it was a hot dog boat. That is a thing. I can just imagine the ingenuity and gumption, let alone the conversation needed, to build oneself a hot dog boat.

"*Honey, I have an idea. And it's going to be good.*"
"*Oh yeah? What is it?*"
"*A hot dog boat.*"
"*A what?*"
"*A hot dog boat! Like a hot dog cart, but on a boat!*"
"*No.*"
"*Oh come on. It's a good idea! We'll make money on it, I promise.*"
"*How much is that going to cost?*"
"*Only $10,000.*"

"!#%$*@!"

"So we're good, right?"

The owner of the Jolly Canuck Riparian Catering Service was overjoyed to see the two of us when we pulled up behind his boat. He had a smile on his face and talked quite animatedly about his venture. He was even happy to oblige us by answering our important questions.

"How long have you been running a hot dog boat?"

"This is my first year."

"Do you sell things besides hot dogs?"

"Nachos and fries, too!"

"Are you actually busy?"

"Absolutely! What can I get you guys?"

I reached into my pocket to grab my wallet, but came up empty. Not wanting it to get wet, I had left it at home!

"Chris! Do you have your wallet?"

He checked his pockets. Empty as well. Unbelievable.

We apologized, thanked him for his time, and pushed off his boat to leave. But that jolly old captain stopped us. He insisted that we take two free hot dogs, complete with fillings, for the paddle home. We happily accepted and all but swallowed them whole.

We drank all our water after eating those hot dogs. Without our wallets, we weren't sure how we'd procure more water at the Forks, so we turned around and headed home. Now travelling with the current, we paddled back in half the time. We lifted the canoe back on our heads at Fraser's Grove Park and trekked back across the six lanes of Henderson Highway to my house.

Ashley didn't believe our incredulous story when we told her that we'd been given two free hot dogs from a hot dog boat. And honestly? I don't blame her.

Two months later, I found out that the Jolly Canuck Riparian Catering Service hadn't secured the proper permits to sell food from the federal docks at the Forks. That summer turned out to be both his first and final foray of selling hot dogs from a boat. When interviewed by the CBC about this oversight, the owner said, "I'm not as jolly as I should be." Not jolly indeed.

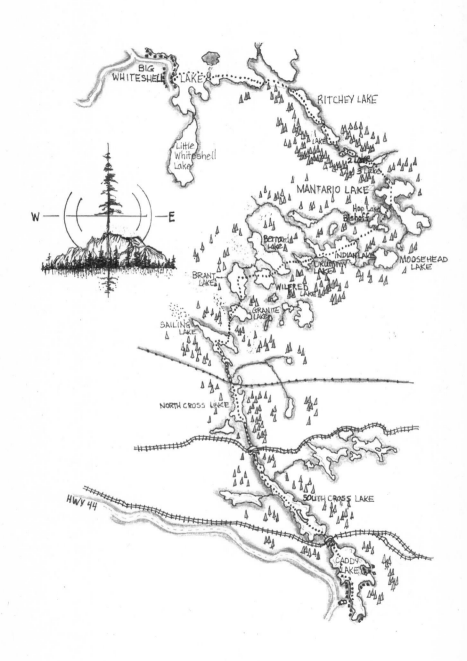

- 4 -

The Lost Hikers

August 2008

I went canoeing with David again. This time, though, we brought along a map. We also added two more people to our paddling party: my sister-in-law Joanne, and our friend Kevin.

We left the old red clunker of a canoe behind for this trip. In its place we were paddling my brand new 17-foot dark green Wenonah Spirit II Royalex canoe. The Spirit II is described by its maker, Wenonah Canoes, as a Swiss Army Knife. "The Spirit II does everything. As efficient as a long-hulled canoe and as maneuverable as a short one, the Spirit II strikes a balance between all good things. Roomy enough for multi-day trips, it's even more stable when loaded." That description is remarkably accurate. There is nowhere I can't take my canoe, and I trust it with my life.

In addition to my Spirit II, we also had Kevin's canoe—an old sixteen-foot Misty River aluminum clunker that belonged to his dad. It was less of a beast than the old red one, but it was still a brute. In any case, it floated and was available at no cost, and thus suited our purposes.

And what were those said purposes? The Mantario Wilderness Zone.

Part of the Whiteshell Provincial Park, the Mantario Wilderness Zone is where hunting, motorized vehicles, and resource development are prohibited. The plan was to tackle nineteen lakes in three days, and carry our gear and canoes for ten kilometres over fifteen portages.

Our trip commenced early on a Wednesday morning in late August, exchanging the longer evenings of July for the fewer bugs of August. With so many portages through bug-breeding grounds, we were constantly grateful for this decision. The first few hours went as expected. North on Caddy Lake, and through the first tunnel. North on South Cross Lake, and through the second tunnel. North on North Cross Lake, a short carry around the dam into Sailing Lake, and then east to the portage into Granite Lake.

We attacked the first portage with enthusiasm. Not wanting to tire ourselves out, we decided to do it in two trips—the first with the canoes, the second with our bags. We finished the 750-metre hike with a sense of triumph and energy to spare. We also realized that, with thirteen more portages to go, making two trips per crossing wasn't feasible. A short paddle on Granite Lake brought us to the Brant Lake portage, and we made a new plan.

David was the strongest, so he'd put on the smallest backpack, hoist the Spirit II onto his shoulders, and start walking.

Joanne, the smallest of us four, wore the largest backpack on her back and strapped the food barrel onto her front. With these bags she was now almost as wide as she was tall, but soldiered on through the bush after David with no troubles.

Kevin and I strapped the life jackets to the middle thwart of his dad's canoe, and lashed the four paddles to the other thwarts with bungee cords. We donned the last two backpacks and tandem-carried his father's trustworthy water rocket down the path. We soon discovered that if we rested the canoe on our backpacks the bulk of its weight transferred down to our hips, thus saving our necks and shoulders from the thwarts digging into them.

The path from Granite Lake to Brant Lake was overgrown in many places and clearly hadn't been used very often in the past few years. This did not bode well for the remaining portages farther into the bush. We were saved by stacks of rocks made years ago by fellow paddlers, as they eventually led us to a straight path through a forest of young willow trees. This led up a ridge, and then we took a hard right down towards Brant Lake. After another hard right around a pine tree and into some reeds, we turned left. Here we found ourselves on the remains of a beaver lodge.

"This can't be the portage landing, can it?"

We scanned the nearby trees and saw the marker hanging from a nearby tree. Three two-by-fours nailed together in a triangle, painted yellow with orange tips. Across the bottom were the words *Granite Lake: 0.5 km*.

"Yup. This is it."

We scouted Brant Lake from the beaver lodge. All we could see were stiff reeds rising four feet above the water, swaying in the wind. A wall of green was standing in our way.

We looked at each other. "What the heck is this? What are we supposed to do?"

Our campsite was waiting for us at the end of the next portage, so we did what any intrepid group of canoeists would do after a full day's paddle—we kept paddling. The portage was on our right, but the reeds stood guard across the entire width of the lake. Not seeing another option, we attacked them head on, pointing our bows straight into the wall.

Thousands of little seeds started raining down on our heads, falling into our canoes. We picked up handfuls from the bottom of our boats. It was wild rice! With Brant Lake a full two portages from where one can drive a motorboat, this rice hadn't been harvested this year.

I glanced around, excited to share my discovery with David and Joanne in the other canoe, but they were gone. The curtain of rice stalks was so thick that we lost sight of each other.

Kevin, paddling stern, turned his paddle into an outrigger, and I stood up. All I could see was rice, shoreline, and sky. Where were David and Joanne?

I called out. "Hey! Where are you guys?"

"Right here!" was the response. I saw nothing but green stalks poking out of the water.

"Raise your paddles!"

The blades of their paddles rose above the vegetation, a mere ten metres away.

"This is ridiculous," Kevin said. "Let's head to shore and look for the portage."

Sticking together this time, we broke free from the claustrophobic press of the rice close to shore. After a few minutes of paddling along the shoreline, we saw the now familiar triangle. *Wilfred Lake: 0.25 km.*

We took the gear out of the boats and flipped them upside down. Bushels of rice fell to the ground. After a few good shakes we had dislodged most of it, so we switched into portage mode and took off toward the campsite.

Kevin was the only one of us who had paddled in the Mantario Wilderness Zone before, having gone all the way to Indian Lake a few years earlier. He suggested that we camp at the end of the portage into Wilfred Lake, and we were not disappointed when we arrived.

The entire campsite was a rock face, sloping gently into the lake. There was one large grassy spot for our tent, a table made of rocks resembling an Inukshuk, and reams of driftwood scattered everywhere.

I couldn't believe it—it was the end of August, and the campsite was littered with firewood! We must have been the only ones there that summer.

"Perfect. Let's make a big fire," David said, as though that was the only logical course of action.

We set up our tent, made supper, cleaned up, and explored some nearby cliffs. We collected all the firewood we could, erecting a large firewood teepee inside the rock ring we had assembled. With the outer layer of wood measuring six feet high, we filled the inner layer with dry brush and branches. The flames licked the night sky.

The next day we paddled across Wilfred Lake, then hiked the 1.25 km to Drummy Lake. After Drummy Lake, it was a 500-metre portage into Indian Lake. We stopped for lunch on a beach, and then it was a 1.75-km portage into Bishoff Lake.

Famed canoeist, naturalist, author, and filmmaker Bill Mason was right. Anybody who says they like to portage is either a liar or crazy.

Apart from the initial triangle signs indicating the beginning and end of each portage, the trail was marked by orange signs with black arrows pointing us in the right direction. We were grateful for these, as we were far enough in the backcountry that we couldn't rely on piles of rocks stacked decades ago by kindhearted souls. Often what we thought was the portage trail was actually an animal trail, winding its way aimlessly to and fro in the bush. Those were often tempting to follow until an orange arrow was found to set us right.

We only paddled a few minutes on Bishoff Lake before we were at the next portage. We looked up at the sign. *Hop Lake: 0.5 km.* At this point in the trip we had become quite convinced that we were portaging more than we were paddling, and all the lakes and portages were starting to blend into one. After Hop Lake, though, was Mantario Lake, our final destination for that day. We knew our day's labour would soon be over.

It was routine for us by now. We jumped out of the canoes, put the backpacks on, hoisted the canoes over our heads, and started walking. David first, Joanne next, with Kevin and me taking up the rear.

The portage into Hop Lake was business as usual—until I fell through a beaver dam.

Without any notice, my right leg broke through a soft spot in an abandoned beaver dam we were walking on, stopping just past my knee. My left knee bent down to compensate, and my arms flexed trying to stop the canoe from landing on my head.

"Whoa, Kyle! Are you alright?" Kevin asked, trying to balance the canoe behind me.

I took a breath to collect myself. "I think so. Let's just put the canoe down and see."

We flipped it over beside us, and I started to look for the best way to extricate my leg from the dam. I was lucky not to have sprained my knee, and thankful that I hadn't dropped the metal canoe on my head.

Joanne had turned around to see the upside-down canoe resting at a funny angle. When she realized that we were okay, she took off to catch up to David. He was silently solo portaging up the trail, unaware of what had just transpired behind him.

Kevin and I dragged the canoe off the beaver dam to prevent any further surprises. We took a few minutes to collect ourselves. I checked the range of motion on my knee and did some little lunges to test how much pressure it could take. When everything checked out fine we lifted the canoe and followed the other two.

Compared to some of the longer portages, 500 metres was a breeze. After a few minutes of walking Kevin and I lifted the canoe, saw the lake up ahead, and made our way down to the water's edge.

We flipped the canoe off our heads and looked around.

We were alone.

"Where are Joanne and David?" I asked.

Kevin turned in a complete circle. "I have no idea."

Unbuckling our bags, we started calling.

"JOANNE! DAVID! WHERE ARE YOU?"

Silence.

I thought about the explicit instructions my father-in-law gave before the trip: Make sure Joanne comes home safely. And here I had gone and lost her in the middle of the Mantario Wilderness Zone. I could feel the panic welling up inside of me.

We kept calling.

"JOANNE! DAVID! WHERE ARE YOU?"

Silence.

"Where could they be, Kevin? It was only a 500-metre portage! Maybe we passed them up there and didn't realize it because we had a canoe over our heads?" I knew this was a stretch, but what else could have happened?

I started running back up the trail, calling, while Kevin went down to the lake to see if they were already there.

"HELLO?!?" I called, my voice getting more and more frantic. "CAN ANYBODY HEAR ME?"

This time, somebody answered.

"Hellloooooo!"

The response came from across the lake. I whipped my head around and started towards the lake. Kevin, already there, perked his head up and looked in the direction of the call.

"Where are you guys?" we yelled across the lake.

"Over here!" they called back. We couldn't see them.

"What are you doing on the other side of the lake? How did you get there?"

"We walked here!"

My panic was turning to anger. I was not in the mood for these kinds of shenanigans. They clearly did not understand how worried we were about them.

"Come to the lake so we can see you!" we yelled.

"No! We're going now!" What was their problem? "Bye!" The whole time we had been talking to a man's voice, but at this point, we heard him laughing, and we heard two distinct female voices laugh with him. Three voices?

"Kevin, are there three people there?"

"Yeah. I heard that too."

"That's not David and Joanne! That must be some other hikers. But where are David and Joanne?"

We heard a branch snap. Our heads turned in tandem. Twenty metres to our left, we saw an overturned green canoe trudging through the forest. We had found David! We looked behind him. Joanne was nowhere in sight.

"David! Where's Joanne?"

He kept walking.

"Hey, David! Where's my sister-in-law?"

Still nothing except David's slow plodding.

"DAVID! WHY AREN'T YOU RESPONDING? WHERE'S JOANNE?" The panic was back.

"I'm right here, Kyle!" About fifteen metres behind him, the bright blue food barrel came into view first.

I exhaled.

Joanne came up to us.

"Where were you guys?" Kevin asked.

"Oh, up there. We didn't know where the trail was, and we ended up following a random animal path. After a few minutes we realized it wasn't the trail, so we headed back. When we saw water, we made straight for it."

"You didn't hear us calling you?"

"Nope. Did you hear us calling you?"

I shook my head.

The sound of branches scraping against plastic alerted us that David was no longer carrying the canoe. He looked around and slowly stretched, then casually reached down to eat some trail mix. His eyes finally met mine, oblivious to the daggers mine were sending.

"David! Why didn't you respond to me when you were so close?" I shouted.

He looked at me, insulted by my question. "Maybe because there was a canoe on my head and I couldn't hear you?"

I sat down and drank some water, frustrated at both my anger and my panic. We combined our stories, rebuilding the last half hour. The beaver dam. The other hikers. The animal trail. All on one of the shortest portages of the trip. What a barely averted disaster.

We paddled across Hop Lake for about ninety seconds, the lake living up to its name. A 250-metre portage crossing the Mantario Hiking trail led into Mantario Lake, where we paddled north for the final stretch of the day.

Mantario Lake was utterly beautiful. Most of the lakes in the Mantario Wilderness Zone are known as "bathtub lakes", as they have steep, rocky shores surrounding their warm, summer water. Deep green pine trees studded the shoreline, accented by the white trunks of birch trees whose crowns were just beginning to turn yellow. Deadfall was everywhere, as the land had never been cleared by loggers or cottagers.

The familiar sound of our canoe hulls crunching the sand signaled that we were done moving for the day. Our campsite was near the portage into Three Lake, a well-used beachy area overlooking some small islands. The site is frequented by both canoeists and hikers, so a picnic table had been constructed there years ago. We set up our tents, collected firewood, made supper, cleaned up, and sat down by the campfire, content with our day's work.

As the sun set, we heard branches breaking in the woods, northeast of our campsite, opposite the lake. The four of us looked at each other. They sounded like footsteps. They kept getting louder, and they weren't slowing down. David made a move for the hatchet.

"Look! A fire!"

We heard their voices before we saw them. Two hikers crashed through the tree line into our campsite. A man and a woman, both in their early twenties with large backpacks on their backs, appeared. The relief on their faces was evident.

They introduced themselves as Kelsey and Tyler.

"We know that sharing campsites isn't ideal, but can we join you for the night?" Kelsey asked.

"Of course," we said. My eyes met David's, and, remembering that time the drunk Frenchmen rescued us, we smiled. We told the hikers to take it easy, set up their tents, and come by the fire to warm up when they were ready. We even had some coffee and chocolate bars to share.

When they finally sat down, we explained our story. Four canoeists, paddling and portaging, heading home tomorrow.

We asked them for their story, and how they came stumbling into our campsite as darkness was approaching. Kelsey and Tyler looked like well-prepared and competent hikers, but almost everything they had done the last two days contradicted their appearance.

They were hiking the Mantario Trail, a 63-kilometre trail through the rugged Canadian Shield, and, sure enough, they didn't have a specific Mantario Trail map. They did have a topographical map and a GPS, though. At the trailhead, Tyler marked the coordinates down for the campsites along the way and planned to trust the little blue arrows marking the hiking trail. This might have been a decent compromise, except that when Tyler turned on the GPS to mark the campsite coordinates, the low-battery light came on. They hoped that if they turned on their GPS, recorded their coordinates, and then turned it off, the low battery wouldn't be a problem because they could just locate themselves on their topographical map. Except that their topographical map was printed in the 1970s, and beaver dams tend to change the ebb and flow of water over forty years. Having a compass along might have helped, but as they were planning to rely on the GPS, they didn't pack one of those either.

But they were young and enthusiastic and had the gear and food all packed, so off they set on the Mantario Hiking Trail. Somewhere alongside Ritchey Lake, they got lost. They were walking along a sparsely treed rock face and missed the trail. There were no blue arrows marking the trail to be found, and while their GPS did a good job of showing them where they were, it did a terrible job of showing them where to go. They figured they'd head due southeast, as that's where the trail went. They never did find the trail again until they stumbled upon us the next day.

They had blazed their own trail, which is never a great idea. Forests are thick, cliffs change plans, and beavers reroute creeks that force kilometres-long detours. When one of those creeks was standing in their way, but wasn't on their topographical map, they shed their clothing and traversed the creek holding their bags high above their heads.

Afraid to check their progress on the GPS for fear of draining the battery, they had only checked their coordinates when they stopped for the night at a random clearing. The GPS showed they had headed more south than east, and were now due west of Mantario Lake.

That night, Kelsey turned on her phone and saw one bar of cell reception. She tried calling home. Someone picked up, but was immediately disconnected. So now, not only were they in a pinch, but their families back home would be worried too.

The next day, they broke camp and decided to head straight east in an effort to find either the hiking trail or Mantario Lake. They started blazing trail again. They wound their way around tree trunks and climbed through the underbrush for hours until they came across our warm campfire. Amazed at their story, all six of us were glad they had found us when they did.

They were about ten kilometres behind where they had hoped to be, making the likelihood of them finishing their trip on time improbable. And since their family was probably already worried about them due to the dropped phone call, we figured that adding an extra day to their hiking plan wouldn't go over too well. We offered to paddle them halfway down Mantario Lake in the morning, to where our portage path crossed the hiking trail. Knowing they'd save a few kilometres of walking, our offer was accepted.

They went to bed shortly after, exhausted and thankful.

Rising early, after eating a quick breakfast together, Kevin and I ferried them to the trail crossing. With four adults and two large backpacks our canoe was sitting low in the calm morning water. The only sound was that of our paddles dipping into the blue. Kelsey and Tyler sat in silence, contemplating what their next twelve hours would entail.

We got to the crossing, found a blue arrow marking the trail, and sent them on their way. By the time Kevin and I got back to camp, David and Joanne were putting the finishing touches on cleaning up the site. We loaded up the canoes again and were off, another day of paddling and portaging, this time with the goal of our warm beds that night.

WHO GOES CANOEING WITH THEIR MOTHER-IN-LAW?

It was a short paddle on Mantario Lake to the portage into Three Lake. A choir of green leopard frogs serenaded us at both ends of the portage.

The paddle on Three Lake was fine, but the portage from Three Lake into Two Lake was less so. We loaded up the canoes as usual, followed the trail marked by the orange arrows, and climbed up the initial incline. At the top, we looked down and saw that the path led straight into a bog filled with stagnant water, bright green algae, and a pungent smell.

We put down the canoes and Kevin followed the path to the edge of the bog. He scanned our side of the bog and as far as he could see in both directions. There was nothing but water.

He slowly searched the other side.

"There! An orange arrow! This bog runs through our trail!"

"Stupid beavers," we commented from the top of the hill.

And so we added an extra little paddle between Three Lake and Two Lake, which shouldn't have been a big deal—except for what greeted us at the end of Two Lake.

"You've got to be kidding me," Joanne groaned as she stared up at the large, steep hill behind the portage marker that read *One Lake: 0.5km*.

Before the trip, I had read that this portage was nicknamed the "Up and Over" portage, because you have to drag your canoe up a hill with a thirty-metre elevation change, and then drag it back down to Two Lake for a fifty-metre elevation change.

Shaking our heads in resigned disbelief, we started climbing. Kevin helped Dave get up the hill with the metal canoe, while Joanne and I carried the Spirit II.

At the top of the hill, we put the canoe down and looked at the water way down below us.

"What are we doing way up here with a canoe?" I asked.

Joanne shook her head slowly in response. "That really is a good question, isn't it?"

Back down we went, towards the next lake. The portage out of One Lake ended with us staring into the wind on the long and narrow Ritchey Lake, white caps smashing against the shore by our feet.

Our communication was automatic at this point. Without a word, we put the canoes in the water, loaded our bags in, and started paddling.

All I remember is that my jelly arms ensured our slow progress, and how my mantra across the entirety of Ritchey Lake was "It's Peanut Butter Jelly Time!" by the Buckwheat Boyz. "Peanut Butter Jelly Time" over and over again. It was horrible.

We made landfall, overjoyed to be off the wind tunnel that was Ritchey Lake, yet full of dread because our last portage into Big Whiteshell Lake was the longest of the whole trip. We read the lettering on the last yellow portage sign we'd see. *Big Whiteshell Lake: 2 km.*

We were right on schedule, and none of us really had much left in the tank to power through this portage, so we did it in two trips. Bags first (thankfully the food barrel was quite empty by this time), a leisurely walk back to Ritchey Lake, and then the canoes.

Since we had been paddling straight into the wind on Ritchey Lake, we thought the wind was coming from the northwest.

We were wrong.

We looked out over Big Whiteshell Lake, knowing that the end of our trip was five kilometres due west, but saw only the whitecaps from the westerly wind blowing right in our faces.

It was "Peanut Butter Jelly Time" again.

Kevin and I did our best to hug the south shore, trying to find some calmer water out of the wind, but David and Joanne stoically paddled straight into the wind, wanting nothing more than for this trip to be over.

Two hours later, it was.

We pulled up to the boat launch and raised our paddles over our heads in weary triumph. We had paddled through the Mantario Wilderness Zone.

Three days, two nights, nineteen lakes, and fifteen portages covering 47 kilometres.

We did it. And we couldn't have been prouder.

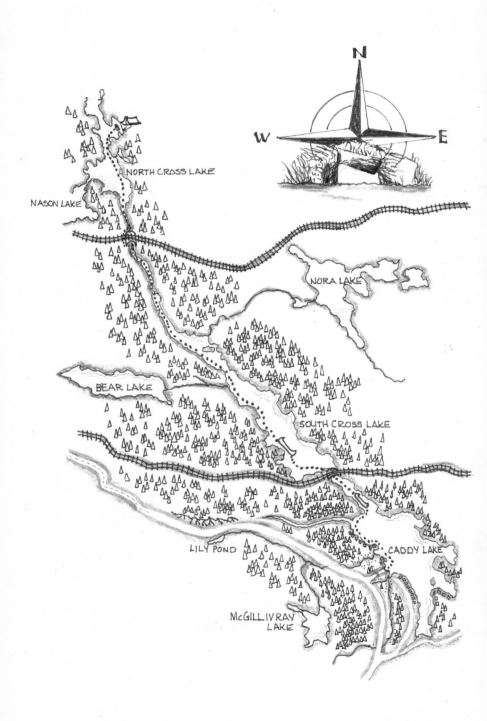

– 5 –

Stuck Between a Rock and a Hard Place

July 2009

The previous summer, our friend Elisabeth took us canoeing for a few days to one of her favourite routes on Dogtooth Lake in Ontario. The next summer, Ashley and I returned the favour. Although our schedules only allowed a one-night trip, the plan was to take her through the Caddy Lake tunnels to our favourite island campsite on North Cross Lake.

As the day of our departure got closer, it started to rain. And rain. And rain some more. Heavy rain, light rain, big drops, little drops, vertical rain, and horizontal rain; it didn't matter. We saw it all.

It had been a snowy winter with a late spring thaw, so the rainwater was joining the remnants of the spring melt. We knew the water levels were going to be high, but how high? And more importantly, would we be able to get through the tunnels?

We'd probably be able to get through the first tunnel without issue, as it had a higher roof. But the second tunnel into North Cross Lake? That was a different story. We could touch the roof of that tunnel with normal water levels, let alone high ones. Would it even be passable?

There was only one way to find out. Plus, if we couldn't get through, there were enough campsites on South Cross Lake that we could just turn around and find somewhere else to camp.

The sun came out just in time for our trip, and the three of us set off in my canoe. Elisabeth is an avid canoeist, so we flipped a coin as to who would paddle stern the first day. I won the toss, and agreed to switch the next day. Ashley was quite content to relax on the sleeping bags in the middle of the boat, her primary concern being to enjoy the sun and the good company. Elisabeth and I were happy to oblige.

After paddling for an hour on Caddy Lake, we cautiously approached the first tunnel, parking ourselves in the reeds along the shore to investigate. We could feel the water tugging at our canoe, wanting to pull us into the tunnel. The water was definitely high, but the tunnel seemed passable.

We tightened our lifejackets, blew our whistle three times to alert any oncoming boats, and set off. I didn't take any strokes, only using my paddle as a rudder. Ashley and Elisabeth both had their arms ready to push us off the walls, but it was unnecessary. We floated through with no problems.

We all agreed that this first tunnel was quite anti-climactic, and looked forward to seeing how we'd fare at the second one.

An hour and a half later, we were peering into its dark mouth. The roof of this tunnel was lower. Much lower. We could see light at the end of the tunnel, 150 metres away, but not very much of it.

We looked at what a portage would entail. Above us was a steep climb to the railway tracks over loose rocks, and we knew there'd be another sharp descent on the other side. We assessed our gear, packed assuming there would be no portages. We saw folding chairs, coolers, and four-litre jugs of water. No way did we want to make multiple trips up and over the rail tracks.

We looked back at the tunnel.

"Let's get closer to see if we can fit," Elisabeth said.

Elisabeth and I back-paddled to keep us in control, letting the water pull us along. As we approached the mouth of the tunnel, Elisabeth laid back and measured.

"Kyle, if you and I lie down, we can make it. But I don't think Ashley can get low enough with all the gear in the middle."

We back-paddled out of the major current. We looked up at the hill again.

"Should we head back to a campsite on South Cross?" I asked.

"I don't know," Elisabeth said. "I really want to see this island campsite you talked about."

Ashley spoke up. "How about this? I'll climb over and meet you two on the other side. If it's just me scrambling up there it shouldn't be too hard. Then you two can duck and get through the tunnel."

We thought about her compromise. "Sounds good to me!" I said. So we dropped her off, promising that we'd be safe.

And safe we were. We again tightened our life jackets and blew our whistle. And then I pointed the bow of the canoe into the tunnel and committed.

Elisabeth and I both flattened out. Not only was it difficult to paddle well while lying down, but our hands on the butt ends of our paddles were scraping the rocky roof. So we switched strategies. Stashing our paddles in the canoe, we grabbed the rocks above us, guiding ourselves towards the light. In the middle of the tunnel the roof dipped down a bit, but we angled ourselves around it without too much effort.

Emerging to daylight, we expected Ashley to be waiting for us, ready to go. But she was nowhere to be found. Trees obstructed our view up and behind us, so we paddled ahead a bit and turned around. We spotted her climbing over the top and starting the descent towards us.

"Come on Ashley! What's taking you so long?" we goaded her.

"Hey! It's really steep up here!" she yelled back. "Besides, these rocks are all loose and I keep sliding down! Plus I'm wearing flip flops!"

She scampered down, far slower than I thought she should be moving, but she was choosing each footstep carefully.

Finally, she got to the bottom. Elisabeth and I paddled up to her, and she hopped in.

"Oh my goodness," she said. "There is no way we can get our canoe and gear up that hill. It's way harder than it looks!"

"Well, that's good to know now," I said. "Because there's no way back to our car tomorrow other than through this tunnel."

Elisabeth was impressed with our island and its tall, straight trees waving in unison to the rhythm of the wind. We marveled at the trees—they looked like dandelions because all the lower tree branches had been broken off for firewood.

Given that it's an island on a well-travelled route, I didn't expect a lot of wood to be available. So Elisabeth joined me in heading to the mainland to collect some wood for the evening's fire. We spotted a large poplar tree recently blown over from a windstorm, and I was imagining all the kindling we could get from the small branches. But the tree was on a steep slope, and the mosquitoes had found us, so I pulled out my hatchet and chopped the tree at its base. We hauled the entire tree straight into the canoe and let the branches hang in the water. We hopped in, trying to not tear our clothes on stray twigs.

We turned the boat around and began paddling back to the island. A few strokes from shore we entered a weedy part of the bay and the canoe ground to a halt. Elisabeth and I dug our paddles into the water as hard as we could, but we didn't move. The overhanging tree branches were enmeshed with seaweed. We were stuck.

"Ashley!" I called. "Help us!"

She dropped what she was doing and bolted to the side of the island closest to us. When she saw us, she examined our predicament.

"Seriously Kyle? You're only ten metres from the island! Why are you calling me like you're in trouble?" she asked, clearly not impressed.

I smiled. "Well, we're actually stuck. Can you throw us a rope?"

"Why don't you just throw the tree overboard?"

"We thought about that. But then our firewood would be all wet, and we'd have to go get some more. And we don't want to do that. Can you just throw us a rope, please?"

Ashley went and got some rope. She tied a small piece of wood to one end and threw it our way. Elisabeth caught it, and Ashley begrudgingly pulled us out of the weeds to open water. Free from our organic chains, we were able to paddle ourselves back to our campsite. And, in our defense, we made a great campfire that night.

The next day our only thought was how to get back past that tunnel. Ashley doubled down on insisting that portaging over it was going to be a ridiculous amount of work and quite unsafe with all the loose rocks. As we paddled towards the tunnel, we hatched a plan.

We'd do what we did the day before. Ashley would climb over the rail tracks again. Knowing that we were now working against the current, Elisabeth and I would paddle up to the mouth of the tunnel and use our hands again to pull us along.

The first part of the plan was executed to perfection. We dropped Ashley off, and she started climbing.

The second part of the plan, though, needed a bit of work. After dropping Ashley off, Elisabeth and I started paddling as hard as we could against the current to get to the tunnel. We progressed a few metres, but the current was too strong and it spit us back out.

We regrouped in an eddy, dropped to our knees, and dug our paddles into the water as hard and fast as we could. We fought the current, making it as far as we did the first time. This time, instead of being pushed back we were able to keep ourselves stationary for a few moments. It was a tie. But not for long.

As soon as our paddling rhythm slowed just a fraction the current got the upper hand and out we went, back to the eddy to regroup.

"Well, this isn't going to work. We just can't make it against that current," Elisabeth said.

I looked up the hill. Ashley was out of sight already, probably on the other side waiting to heckle us.

"Let's get as far as we can once more," I said. "Only this time I'll jump out on shore and pull us to the tunnel entrance."

So we dropped to our knees again for round three against the current. We made it to the same spot as previous attempts, but this time Elisabeth pointed the bow towards shore. I was able to grab the branches of a small tree and haul us towards the trunk. I clung to the tree trunk long enough for Elisabeth to grab the shore and anchor the boat. I jumped out, holding a rope we had tied to the bow.

I dragged the canoe against the current, inch by inch. Elisabeth used her paddle as an outrigger, keeping us stable. I walked us as close to the rock face as I could. When Elisabeth got ahold of a tree near her, I grasped a handhold inside the tunnel and jumped back into the canoe.

But one hand on the rock wasn't enough to hold us against the power of the current. My grasp gave way and we were pushed back into the eddy. Round three went to the current.

We tried again. We paddled up and I grabbed the tree, hopped out, and dragged us forward. Elisabeth clutched the same tree trunk, but instead of having one hand grab the inside of the tunnel, I worked hard to have two hands clasp some rock. I awkwardly held the rope in my second hand, now also grabbing a lip of rock, and jumped back into the canoe. This time, I was able to hold on.

I pulled the canoe forward, handhold by handhold, slowly leaning backward as the roof got lower and lower. I was amazed at the power of the current, and thankful when I had pulled us the necessary three metres for Elisabeth to join my bouldering.

And so we went. The two of us lying down, pulling ourselves forward along the roof of the tunnel. In the commotion to get as far as we did, we had forgotten to put on our headlamps. We were crawling along blind to what was just half a metre above us.

We were making decent progress until the middle of the tunnel, where the roof sloped downward. I felt the lower rocks with my hands and kept pulling us along. But when my chest was just past the lowest part of the ceiling, my hands slipped. The current started to push us backwards. We grasped for handholds, but before we could locate good ones, my lifejacket jammed up against the low-hanging rock. The canoe stopped moving.

As the current tried to push us back, it also tried to push us down. The rock I was stuck against sloped downwards, so every centimetre we moved back was also a centimetre we moved deeper into the water. The problem with this was that as the boat was pushed deeper it displaced more water than it had to, and thus pushed back up with even more force. And the major contact point was my lifejacket. I was literally caught between a rock and a hard place. In the dark, no less.

Equilibrium was finally reached. We stopped moving either back or down. But I couldn't move my torso. I called back to Elisabeth.

"I'm stuck here! We need to push down against the ceiling to get my lifejacket off this rock. Then we need to shift ourselves a bit to the left to avoid this low point and pull ourselves forward. Got it?"

"Push down. Move left. Pull forward. Got it!"

And we did. We pushed the canoe lower into the water to get my life jacket unhooked. We floated back a metre but found new handholds to steady ourselves. We shifted to the left and resumed our forward progress. The low rock passed just to the right of my face. *This is nuts*, I thought. *We never should have come through here yesterday.*

Elisabeth called to me when she was past the low-hanging rock, and we continued crawling through the rest of the tunnel. We made it to the other side, our eyes squinting in the bright light after the blind dark of the tunnel.

We sat up as soon as we could and paddled off to the side so the current wouldn't pull us back into the tunnel. We turned around and looked above the mouth of the tunnel. Sure enough, Ashley was there waiting for us.

"What took you two so long?" she asked, a big grin on her face.

I smiled back. "You have no idea, Ashley. You have no idea."

She took a picture of Elisabeth and me coming out of that tunnel. It's beautiful. The water is calm, reflecting the sky and clouds, and we're framed by green leaves and pine trees. We're paddling hard against the current; our canoe is giving off a small wake.

We love that picture. And even though it's Elisabeth in the canoe and not Ashley, we have it framed in our living room. It paints a tranquil picture, every canoeist's dream, oblivious to the terror we had experienced only moments before.

Maybe that's the allure of canoeing—you work hard, take calculated risks, and adjust on the fly when things don't work out as planned. And, in the end, the glimpses of heaven are worth it.

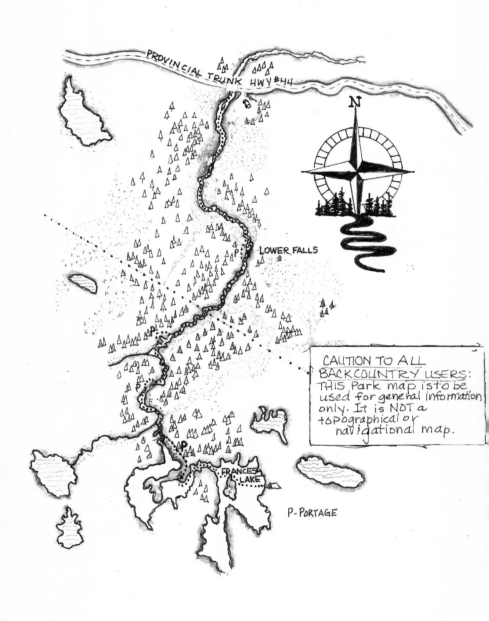

– 6 –

Who Goes Canoeing with Their Mother-in-Law?

July 2009

"What? You went on a canoe trip with your mother-in-law? Who does that?"

These are the questions I receive when I tell people I went canoeing with my mother-in-law. Questions, and plenty of strange looks.

Who goes canoeing with their mother-in-law? Well, I do. And we had a great time, too.

My mother-in-law Patti, Ashley, and I had gone for a day paddle on Hanson's Creek the previous summer. We had a free afternoon, so we set the casual goal of paddling to some power lines marked on the map. I had counted two-and-a-half map squares from the highway to the power lines, and estimated that would be a five-kilometre round trip. We could do that in about ninety minutes.

The highlight of this day's paddle was stumbling upon a wolf eating a deer. We saw the wolf look up from the river and start running along the reedy river bank. It climbed up a hill and stared at us as we paddled by. When we got to where it had been, sure enough, there was a deer carcass stuck in the water along the shore. Large chunks of it were missing.

The lowlight of this day's trip was that, after an hour of paddling, we still couldn't even see the powerlines. And the path further upstream was blocked by a small waterfall that would have to be portaged.

Something wasn't right. We couldn't be lost, as there were no navigable rivers flowing in or out of the creek. We pulled out our map to confirm where we were.

The first thing we saw were the words "Lower Falls."

Ashley looked at me and asked if I recalled seeing that there. Nope. I had not.

Oops.

In my defense, there was no capital "P" beside it for "portage." There were other portages marked on the route with a "P," though. How could the cartographer remember to mark "Lower Falls" but forget to put a "P" beside it?

Secondly, we looked at the scale of the map and discovered that each square was not one kilometre, as we had expected, but rather one mile. We were expecting a two-a-half-kilometre paddle to the power lines. In reality, it was a four-kilometre paddle.

Oops.

Thirdly, the river meandered significantly. All those curves added to the overall distance. As the crow flies, the power lines were four kilometres from our car. As the canoe paddles, they were seven.

Oops.

Now frustrated at my map-reading skills, Ashley snatched the map to take a look for herself. She groaned. "Kyle, this isn't a backcountry map. This is just the interpretive map! It literally says "CAUTION TO ALL BACKCOUNTRY USERS: This map is to be used for general information only. It is not a topographical or navigational map. It should be referred to only in conjunction with topographical and/or navigational maps." Shaking her head, she rolled her eyes at my stupidity.

Oops.

We turned around and headed back home, a little sad that we didn't make it to our goal, but still grateful for quality time on the water. Plus, it's not every day you startle a wolf who stares at you as you paddle by.

The following summer I was determined to not only get to the power lines but to paddle all the way up Hanson's Creek to Frances Lake.

I checked with Ashley, and she wasn't too thrilled. She hadn't grown up canoeing and was still following the lead of her father, Doug, who claimed that if there wasn't a motor on the back of the boat he wasn't going. I assumed correctly that he wouldn't be joining me either.

So I asked my mother-in-law to join me. Her immediate answer? Yes!

As we were making a menu plan and packing our gear, Doug made sure to do his part to ensure a successful trip. He pestered us with questions about safety, timing, gear, and emergency protocols.

And he supplied us with the official Whiteshell Backcountry Map. No more interpretive map for us! We were set.

I left Ashley with her dad, and I set off with her mom for a two-day trip.

The first part was just like the previous year's excursion, albeit this time we didn't see a wolf. After two hours of paddling we came up to the now-anticipated waterfall. Interestingly, the cartographers of the up-to-date backcountry map had neglected to mark anything down about these falls. It was just a short portage one-hundred metres over some rocks, so we started carrying our gear without thinking twice.

Unfortunately, what came next was wholly unexpected. We looked out over Hanson's Creek on the upstream side of the falls, and all we saw was marsh, reeds, and a small creek slithering through it like a snake.

"This can't be it," I said. But when we looked down at the water, we saw the current crawling towards the waterfall. This must be the way. So into the marsh we went.

The creek was similar to a switchback road on a mountainside. You go up a bit and then turn one-hundred-and-eighty degrees. Then you go a bit further and turn again. Over and over and over again.

We did this for at least an hour. But our spirits were still high, and we were determined to get to the lake. At the end of the marshy areas, there was another portage! We looked at our map and again saw nothing marked. So much for the official backcountry map.

Lucky for us, this portage was a shorter distance. Unlucky for us, the next bit of creek was almost identical to the previous one—marsh, reeds, and switchbacks. The only difference was this part of the creek was even narrower.

Onwards we went, a bit incredulous at both the deception of the map and the route itself. Plus it was here that the mosquitoes and blackflies found us. They hid from the wind beneath the gunwales and feasted on our ankles. In response, our legs glistened with bug spray.

Some of the switchbacks were so extreme that all I could do was make a hard C-stroke on the same side that Patti was paddling on. This turned the canoe sharply. Then my next few strokes would be reverse strokes on the opposite side of Patti. This would turn us around completely. And then I'd tell Patti to switch sides to where I was paddling, and I'd make a hard C-stroke, repeating the process. There was no forward momentum on that creek, having to start and stop so often. But there was a fair amount of grunting and groaning.

And then there were the beaver dams—all nine of them. We'd see one in the distance and paddle as hard as we could. I'd yell, "Ramming speed!" and ram the hull up the dam as far as it would go, and then climb out onto the mud and branches to lift the canoe the rest of the way over.

Wouldn't you know, after this marshy part of the river filled with beaver dams, there was another portage! Also unmarked on our maps. This one was again, short, but this was getting ridiculous. How could there be three unmarked portages on a backcountry map? I now remembered the "P's" on the interpretive map, wishing we had brought that one along as well.

After this portage we were greeted by yet more marsh. Thankfully, the stream was starting to straighten out slightly so we didn't lose all of our momentum on every turn. But there were at least three more beaver dams.

Throughout all of this, Patti was a superstar. She was in her early fifties and hadn't gone on a lot of canoe trips before. But here she was, rocking the paddling and portaging like a grizzled veteran. She started anticipating the upcoming curves and adjusted her strokes. She kept her paddling rhythm similar to mine. And, when there was a portage, if she wasn't helping me with the canoe, she was carrying most of the other gear so I could solo-carry the canoe. Plus, she was optimistic and encouraging, necessary on any trip, but especially on this challenging route. Patti was the perfect paddling partner.

It had now been four hours since we had started and after two more beaver dams, the marshy, meandering creek ended and we portaged once more. Again, the backcountry map didn't mark this portage. What good was this backcountry map?

This portage was quite a bit longer than the others and ended at an amazing campsite on Frances Lake. It started in typical rocky Canadian Shield, with the trail being surrounded by dry lichen and straggly pine trees. Near the end of the portage, the topography changed into a stand of tall pine trees, all of them having no branches except for their green Christmas-tree tops. The trees swayed with the wind in unison, sporadically letting sunlight peek through to the needle-covered floor. The only noise other than our footsteps was the creaking tree trunks. In all my travels, I have yet to find a place similar to this.

And then there was Frances Lake. We got there in the late afternoon and had a gentle wind at our backs as we paddled east to the other campsite on the lake. Despite being a mere six kilometres from the bustling resort town of Falcon Lake, this lake clearly did not get many visitors. With the exception of the two campsites, there was no evidence of humans.

We paddled through narrow channels and under cliffs, taking in everything. We arrived at our campsite, a rocky ledge jutting into the water, surrounded by trees. The fire pit was overturned, off to the side. We think the last guest was a curious and hungry bear. But, other than that, it was just the two of us: me and my mother-in-law.

We had planned a simple supper of hot dogs and canned beans. After cleaning up we sat on the rock ledge and watched the sun go down. We went to bed tired after a hard day's work, but content with what we had accomplished.

I usually remember most of my canoe trips, but the next day of this trip is mostly a blank. I guess we made it back to the car by paddling through the meandering marsh, only this time abetted by the current. We must have gone back over all those beaver dams, the current increasing our top "Ramming speed!"

But honestly, there are only three details from this day that I remember with any clarity.

I remember that Patti woke up sick. She unzipped the tent and threw up in the bush. When you're on a two-person canoe trip with hours of paddling and portaging ahead of you, being sick is less than ideal. But we did it. I give so much credit to her for powering through while being so sick.

The second detail I remember is that I committed one of the most heinous, unforgivable sins of canoe trips: I had forgotten the coffee. I unpacked the percolator and filled it with water. I put it on the campfire grill, ready to go, and looked for the coffee grinds. After emptying the entire food bag twice and not finding any coffee, I sat down and put my head in my hands. I was only able to mutter one word to my faithful paddling partner.

"Sorry."

And like a mother hen that gathers her chicks under her wing, she sat down by my side and wrapped her arm around me. "It's okay. We'll get through this."

I turned to her and raised my eyebrows. "Really? How?"

"Well, I'm sick, so I wouldn't be drinking coffee anyways. And I have one painkiller left in the first aid kit. You can have it if you get a headache."

My heart swelled with love at the gift of my mother-in-law. Given the circumstances, we broke camp as soon as possible and set off.

And the third detail I remember? When we pulled up to the landing spot, I saw a large brown sign in between the water and our car. What did that sign say? Why hadn't we seen it the day before? Had we walked right past it?

We pulled the canoe out of the water and walked around the sign. In big white letters we read:

Frances Lake Canoe Route
To Frances Lake – 13 km
Paddling Time – 6 hrs
Portages – 4
To campsite – 18 km

We couldn't believe it. After cursing out cartographers for not marking the portages, we had walked right past the information we were looking for—and we had both missed it.

We laughed at ourselves. What else could we do? We had successfully paddled up Hanson's Creek to Frances Lake, and back again. And we both had a great time doing it.

A few years later, somebody asked me a question.

"So, now that you've done it, would you go canoeing with your mother-in-law again?"

Without any hesitation, my answer was crystal clear. "Absolutely."

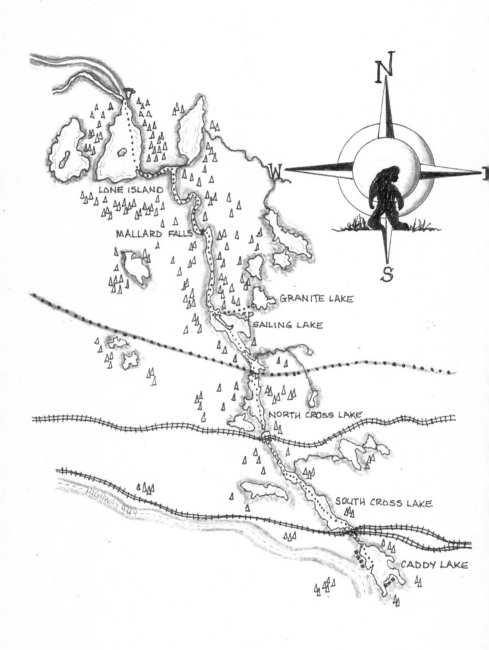

– 7 –

Bigfoot is Real

August 2010

Some people think Bigfoot isn't real. Others disagree and think the animal is elusive. The other day someone saw me wearing my Sasquatch socks and Sasquatch scarf and asked, "Do you actually believe Bigfoot exists?"

There have been hundreds of documented Bigfoot encounters and experiences across North America, with several dozen in Manitoba alone. Those stories are recorded on the Bigfoot Field Researchers Organization's website. There, Bigfoot encounters are classified into three categories:

Class A: A Bigfoot was sighted.

Class B: Evidence that a Bigfoot is in close proximity, such as howls, grunts, smell, or sticks being broken.

Class C: Evidence that a Bigfoot was there, such as a footprint or unidentified hair.

In the Whiteshell Provincial Park, there is a fascinating story of an outdoorsman who repeatedly had Class B and C encounters.

In the mid-nineties, he spent significant time deep in the forest near Big Whiteshell Lake. One night he was in his tent and was hassled by a very loud animal that he couldn't identify. Sticks were breaking around his tent, there was growling, and the animal wouldn't run away when he shouted or threw rocks at it.

A typical Class B encounter.

Two years later, he was again hiking in the same area. On the trail, he was startled by unidentified growling similar to what he had heard previously. And on his way back he almost stepped on a dead grouse with its wings tucked in. It was still warm. He was confident that the grouse wasn't there on his hike in. Perhaps the Sasquatch was trying to connect with him, and the grouse was a food offering of sorts.

Another Class B encounter.

Then, in 2004, he was leading another hike close to the previous ones. He asked his group if they wanted to see if there was a Bigfoot nearby. They said yes, of course. He told them of the legend where, if you strike a large branch three times against a dead pine tree, a Sasquatch in the area will respond in kind. He picked up a branch and struck a tree three times.

They waited in silence. There was no answer.

Except for the dead grouse a few hundred metres away, still warm, with its wings tucked in.

A Class B encounter.

He returned to the site two and a half months later, and the dead pine tree that he had struck was now smashed to pieces.

A Class C encounter.

All these Bigfoot encounters were a mere fifteen kilometres from where we would be canoeing! I was back to paddling the Caddy Lake tunnels route, this time with eleven teenagers and university students, two fathers, Werner and Mel, and my friend James. The fifteen of us knew each other well, and, having been on canoe trips before, we were a confident and gregarious bunch.

We camped on the usual island in North Cross Lake the first night, surrounded by towering pines that swayed in the wind. It was late August. The sun set early and we gathered around the campfire for warmth. I asked the group, "Do you want to see if there's a Bigfoot nearby?"

The jeers and taunts started immediately.

"There's no such thing as a Bigfoot, Kyle."

"You're full of it."

"Bigfoot doesn't exist!"

"Hey everyone! Kyle thinks he can talk to a Bigfoot!"

After the initial skepticism and sass, it quieted down. I told them the story of the Class B and Class C encounters just a few kilometres away from us. I held their attention as I regaled them with stories of the dead grouses on the path, of being watched from the edge of the forest, and communicating with any nearby Sasquatch by smashing a branch against a dead pine tree.

When I was done, I picked up a piece of firewood and asked if we should call any Sasquatch that might be nearby.

They laughed. More AT me than WITH me. But said yes.

I struck a nearby tree three times with the branch.

KNOCK. KNOCK. KNOCK.

We listened to the echo bounce around the lake.

Then silence.

"Yeah, whatever Kyle."

"Like I said, there's no such thing as a Bigfoot."

"Okay, fine. But imagine that we find a dead grouse on the path tomorrow. What would you do? Or, let's pretend that you actually saw a Bigfoot, but didn't get a picture or video. Would you tell anyone? Would you open yourself up to that kind of criticism?

Or would you only tell your family and friends? Or maybe not even them, lest you be that crazy cousin that saw a Sasquatch years ago?"

"Sure thing, Kyle. We're going to bed."

"Wake us up when you see one."

In our tents we listened to the silent forest for anything that might resemble knocks on wood. Nothing. The kids fell asleep confident that our campsite wouldn't be visited by a Bigfoot that night.

The next day we set off north into Sailing Lake. One of the girls in university had never caught a fish before, so I made sure she was in my canoe to help her catch a fish. She cast a few times and felt a tug! She had a small northern pike on her line, but, alas, wasn't able to get it into the boat due to her simultaneous and equal levels of excitement and panic. Although she did admit to me after that maybe it was a good thing it got away, as she was quite scared to hold a wriggling fish. "Plus," she said. "My mom caught her first fish a couple of years ago and she was so excited she peed her pants. I'd rather not follow in her footsteps."

On Sailing Lake, you have a choice of two routes to take.

Paddling north on Sailing Lake will bring you to a small river, then Mallard Lake, and then Mallard Falls. A short portage around the falls will bring you to the Whiteshell River and then on to Lone Island Lake.

Paddling east on Sailing Lake will bring you to a 750-metre portage into Granite Lake and the start of the Mantario Wilderness Zone.

On this trip we had planned for minimal portaging, exhibited by packing a cooler of milk and cookies for the boat ride. But six of us wanted to take a look at Granite Lake, so we split off from the rest of the group and agreed to meet them on the way up to Mallard Falls.

We pulled our canoes up to the shore line at the portage sign and started walking up through thick bush, and then back down to where rays of light poked through the pine branches. Soft ferns were the only underbrush.

Near the end of the trail, Arvid and Adrian, the two in the lead, stopped. Arvid raised one palm up to halt us, and with his other hand he put a finger to his lips. He beckoned me closer and pointed into the underbrush, whispering, "Kyle, what are those?"

I looked where they were pointing, squinting my eyes and waiting for movement. I saw one bird move its head sideways, another jumped onto a log, a third step forward towards us. We had stumbled upon three ruffed grouse.

"Grouse!" I whispered back.

My heartbeat quickened. Somebody asked, "What should we do?"

The correct answer was, "Nothing. Let's just watch them for a bit."

The incorrect answer was what somebody yelled. "GET 'EM!"

James yelled "No, don't!" But it was too late. The thrill of the hunt had seized us. A stick was thrown like a spear towards the grouse. It missed. Another stick hurtled horizontally through the air like a helicopter blade. That branch hit the grouse that was perched on the log, knocking it off in a kerfuffle of feathers.

"We hit one!" someone shouted. They ran to the grouse, picked it up by its feet, and carried it back to the portage trail triumphantly. "Now what we do we do?"

"Great. Now we have to eat it," said James, the seasoned chef of our crew.

"Or," I chimed in, my mind exploding with possibilities. "We could bring it back with us and convince the rest of our group that there's a Bigfoot nearby!" I started imagining the reaction of the teenage boys if they woke up to a dead grouse at their tent door.

The six of us took a vow of secrecy, and started walking back to our canoes. I carried the grouse by its legs, reminding myself to wash its blood off my leg before getting back into the boat. Once there, I found an empty bread bag, threw the grouse into it, and stashed it under my seat.

We turned our canoes north to meet up with the rest of our crew.

We caught up to the canoe being paddled by Sarah and her father, Werner, who were enjoying the tranquility of the river just north of Sailing Lake.

"How was the walk to Granite?" Sarah asked.

"Ah, fine. Not much to speak of, other than the trees, the rocks, and the water. Pretty generic. But it was good to stretch our legs. How are things here?"

"Oh, you know," she smiled "Not much, other than we saw a Bigfoot over there on that hill, smashing a branch against a dead pine tree. It looked at us and walked away."

I laughed. Outwardly at her joke, but inwardly for the chaos that would be unleashed later on that night. My paddling partner and I pulled ahead of their canoe, barely able to contain our snickers.

We paddled through the recently harvested wild rice field of Mallard Lake and began lugging our gear up the slope to make camp at Mallard Falls. On the way up I noticed the bear feces on the path. I put down my bag, found a stick and poked it. Not fresh, but not old either. I picked up my bag and kept walking.

I saw more bear feces. And more. And more. Feces everywhere.

We picked a clearing under the old oak trees, kicking away small piles of bear feces to make room for our tents. I looked nervously at my other leaders.

"There's a bear living here."

"Yeah, but the next campsite ahead of us is two hours away on Lone Island Lake. And we don't really want to paddle an hour back to Sailing Lake, do we?"

WHO GOES CANOEING WITH THEIR MOTHER-IN-LAW?

"Black bears are scared of people. There's fifteen of us, and we're certainly not quiet. It'll leave us alone."

With a bear hanging around I knew that the original plan for the grouse had to change. Leaving a dead grouse near someone's tent is one thing. But giving a free meal to a bear next to someone's tent is another.

We hauled up the rest of the gear and flipped the canoes on top of each other halfway down the portage. Starting our portage the next day already halfway done was enticing. I brought the grouse to camp, hidden in between two other bags, and, when I walked by the overturned canoes, I threw it underneath mine with the intention of retrieving it later.

After supper, two of the kids came running back from exploring the waterfall. They had seen a massive snapping turtle sunning itself in the day's last bit of sunlight. They invited us to come and see, so we left our dessert for later and joined them.

They were right. The turtle was amazing; a living dinosaur. We enjoyed it until somebody got too close and it swam away. We spent a fair amount of time jumping around the waterfall, looking for dry places to step and new vantage points to take it all in. After a while, three of the schemers, plus Sarah, started heading back to camp. I saw my opportunity.

I let the four of them walk ahead of me, just a bit, and glanced behind me. The path was empty. As we walked past the overturned canoes, I reached down underneath my canoe, pulled the grouse out of the bag, and placed it on the path where everyone would see it. Then I went back to the campfire with Sarah and the three other conspirators, and waited.

I didn't have to wait long. Blood-curdling screams split the air. More screams followed. Then shouts. Confusion. Yelling. And then cutting through the chaos was John's voice.

"KYLE! HE DID THIS! THIS IS ALL KYLE'S IDEA! HE SET US UP!"

I was grateful for the waning light—it hid the edges of my lips, which were curling up into an uncontainable smile.

The rest of the group descended upon the five of us already at the fire.

John was furious. "Kyle! This was you!"

"John, how could it have been me? I was with you at the waterfall, and then I came back with Sarah, Becca, Sam, and James. They saw me! I didn't leave! Right, Sarah?"

"He's right, John. He was with us the whole time," Sarah said. I breathed a sigh of relief. Sarah was a straight shooter. I had my alibi.

"So," I asked, "Do you still think I'm stupid for believing in Bigfoot? Clearly there's one here, trying to communicate with us. He or she even left us a food offering!"

"No. That's ridiculous," Paul piped in. "There's no such thing as Bigfoot. Kyle, you brought this bird from home."

"Yes, Paul. You're right. I have a freezer filled with ruffed grouse that I keep just for the purpose of lugging on canoe trips to scare people," I replied sarcastically, shaking my head. "No, I did not bring this bird from home!"

"Well, there's something going on here, guys, and we need to figure out what it is," Phil said.

"We should probably pack up camp and move," Julie chimed in, her speech faster than usual. "We can paddle in the dark. We can do it. We'll use our headlamps, and just set up at the next camp."

"Lone Island Lake is hours away, Julie. Sailing Lake is an hour back. I don't think we're packing up, paddling, and resetting up camp in the dark," I said.

"Is the grouse still there?" Phil asked. "Maybe there's a clue."

Andrew, John, Phil, and I traipsed down the path to the grouse. It was dark at this point, and our headlamps only illuminated a few feet in front of us. The boys were nervous about the grouse and the Bigfoot, and I pretended to be. But we were all nervous about the bear.

When we got to the grouse, Phil smiled. "Okay. This is where I froze and just pointed at the grouse. And then Ryan came up behind me and screamed."

"Wait a second. Was that Ryan screaming?" I asked, laughing at how those boys were driven to terror because of me.

Phil laughed and nodded.

Andrew and John were examining the ground around the grouse, but couldn't find any footprints that didn't belong to us.

"Check if the bird is still warm."

They picked it up.

"Nope. But it's been here for at least thirty minutes already. How long would it take for a dead bird to lose its heat?"

None of us knew the answer. But we all agreed that the grouse shouldn't stay on the path as bear dinner, so we took it back to the campfire.

We sat down again. I drew back from the light of the fire, picked up a small branch, and threw it into the forest. Everybody jumped.

"What was that?" Ryan asked.

"We really can't stay here," Julie said.

"I agree," I said.

"No Kyle! You can't play us like this!" John jumped in. "I know you did it. The grouse must be a fake!"

"A what?"

"You heard me. It's a fake bird. You brought a fake bird on this trip and had this whole thing planned."

"It's not a fake bird, John! Look!" I held up the bird over the fire and cut off a wing with my Buck knife. Blood dripped into the fire.

"Oh, come on Kyle. We can't have a dead, bleeding bird by our campsite," James scolded me. "Go throw it in the waterfall."

I walked down to the water with Paul and John. I gave them one more chance to investigate the bird's authenticity. They agreed the bird was real. I tossed it into the rushing water.

We walked back up to the campsite.

John sat down. "The bird's a fake."

I threw my arms up in the air. "Come on, John! Down there you just agreed it was real!"

He shook his head.

"Okay, let's look at this logically," I said. "There was a grouse on the path. That much we know. How did it get there? Option A: It was walking along and, after the five of us passed by the canoes, it died right there on the path."

Phil laughed. "No. That did not happen. That bird did not just die on the trail."

"Okay. Option B. Somebody put it there. But nobody's been alone this whole evening, and Sarah has already vouched for my presence. Right, Sarah?"

"Right," Sarah said.

"Plus, while I'm honoured that you think I have a freezer full of grouse for this exact moment, that is simply not true."

"What's option C?" Becca asked.

"That a Bigfoot left it there."

A minor uproar resulted. "No way!" "That's simply not true!" "Sasquatch aren't real!"

"Well," I reasoned. "What other options are there?"

"Like I said, it was a fake bird," John insisted.

"John, we've been over this. You saw it bleed! Right, Paul?"

"Yup."

"And besides. Where would I buy a fake bird that bleeds?"

"The internet. eBay. You can buy anything there."

"You think I can buy a fake grouse that actually bleeds on eBay?"

"Yes."

"Fine. Let's do this. I'm going to call Scott and see if you can buy a dead grouse on eBay." I flipped open my phone and saw that I had one bar of service. "Ryan, is your brother at home?"

"Should be."

I dialed his number.

"Hello, Scott! It's Kyle here."

"Hi, Kyle."

"We have a question for you. Can you buy a grouse on eBay?"

No answer.

"Hello? Scott? Can you hear me?"

Silence.

"Hellloooo?! Argh. I lost him."

This was a dumb move on my part, as one of the worst things you can do when camping is call someone but not get a clear signal. They're going to assume the worst and send help. Mortified at my stupidity, I tried calling again, but the signal was too weak. I wondered whether or not we'd be greeted by a search and rescue team in the morning because of my mistake.

My phone beeped. I opened it up and saw a text message from Scott.

"Tomorrow's weather: +23, Sunny, Wind 15km/h from the SE. - Scott"

We all snickered. Scott was an amateur meteorologist at the time, so of course he assumed we wanted a weather update.

I texted him back. "Can you buy a real, dead grouse on eBay?"

He replied shortly. "Found a real, stuffed pheasant in the UK for 25 Pounds, but that's it."

"See John! You can't buy a real, dead grouse on eBay! It was a Sasquatch!"

The accusations went on for a bit longer. Every now and again I'd throw a branch in the forest to scare us, and then someone would wonder out loud if staying the night was the right decision.

Until, finally, James jumped in. "Okay Kyle. It's been over an hour. It's time to tell them."

Oh, how we laughed and laughed. Well, most of us laughed. Some of us were angry, but that soon gave way to a strange sort of respect for how the evening played out.

Phil said, "You got us good, Kyle. You got us good."

That night in our tents, I was with James and the dads.

"Well, we have some scared kids in those tents over there," said Werner.

"Kyle, I didn't know how you got the grouse there, but somehow I knew it was you," Mel added. "I knew it was you because if it WASN'T you, you would have been a lot more scared than you actually were."

Mel was not wrong. If I believed in Bigfoot, and looked up stories on the internet about dead grouses on paths, and then actually found a dead grouse on the path, you'd better believe I'd be terrified of meeting a Sasquatch in the forest.

That night, we gave each tent of kids a whistle and a paddle in case the bear visited. But it ended up being a quiet, peaceful night, for which we were all grateful.

Except in the morning.

Arvid stumbled out of his tent to take a pee, looked up, and saw a bear standing on its hind legs, looking right at him. He froze, still peeing, staring at the bear. The bear watched him for a bit, and then ambled away.

After paddling a few hours in +23°C sunny weather with a gentle 15 km/h wind from the southeast, we were picked up at the boat launch on Lone Island Lake.

We will never forget the canoe trip where we had a Class B encounter with Bigfoot.

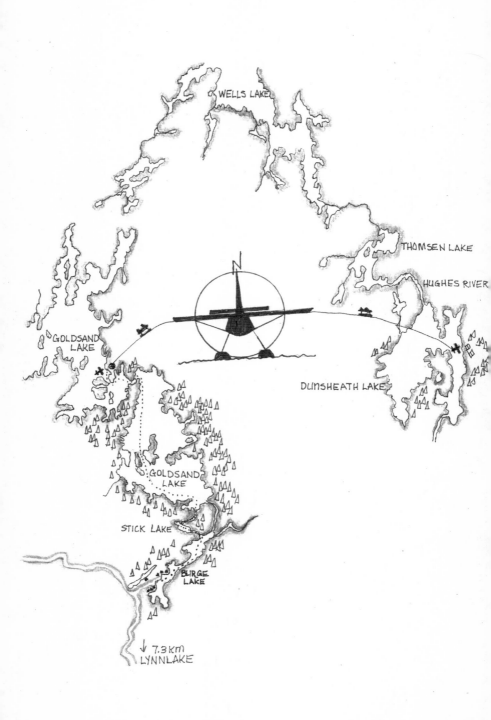

– 8 –

But I Saw a Waggle!

July 2013

Dunsheath Lake is a lake 25 kilometres northeast of the town of Lynn Lake in northern Manitoba. Because of the long winters and brief summers, the trees are short in stature and sparse in density. Hence why the land is historically known as the Land of the Little Sticks.

Despite its relative proximity to Lynn Lake, there is no road to Dunsheath Lake. The muskeg, swamps, lakes, and creeks make walking—let alone driving there—impossible. There are only three ways to get to Dunsheath: You either fly in a floatplane, drive a snowmobile, or paddle a canoe. In July of 2013, I tried to paddle a canoe there.

In 1973, a Cree trapper named Theophile Colomb sold his cabin on Dunsheath Lake to Abe Enns and Russ Broadhead. Abe was the Lynn Lake town administrator, and Russ was a local pastor. After one year of joint ownership, Russ sold his half of the cabin to Abe. Abe made sure to visit as often as he could, flying up with a variety of friends and family. But his favourite times were when he could go up with his wife, Anne, and their four kids, Bev, Esther, Douglas, and Gerald.

In 1987, two of his children, Douglas and Gerald, purchased the cabin from Abe and kept the tradition alive by flying in their families. Doug, his wife, Patti, and their kids, Ashley, Joanne, and Caitlin, would fly up as often as they could. And when the girls started dating, the boyfriends were invited. And when the girls started getting married, the husbands were invited. And that's where we pick up the story.

Joanne married Warren, Caitlin married Jason, and Ashley married me. After years of flying up to Dunsheath Lake together, the in-laws hatched a plan to paddle there the following summer. We'd start at Burge Lake, portage into Goldsand Lake, portage into Wells Lake, paddle into Thomsen Lake, and then down the Hughes River into Dunsheath Lake. After three days of paddling, we'd cruise into Dunsheath Lake where we'd meet the rest of our family who had flown in the day before: Doug and Patti, Ashley, and our two kids, Arianna and Zach, Joanne with her and Warren's kid, Xavier, and Caitlin. We'd end the canoe trip with a week of vacation together at Dunsheath Lake.

The only problem with our enthusiastic group of in-laws was that we were an odd-numbered enthusiastic group of in-laws.

Paddling with an odd number of people is less than ideal. We hemmed and hawed about what we should do. Paddle three in a canoe? Get a kayak? Bring a friend? We were discussing this conundrum at a family gathering at Christmas when Caitlin spoke up.

"Guys, Jason has been bugging me to come. He thinks it'll be good quality time for both the two of us, and with my brothers-in-law. So I guess I'm in. I'll paddle to Dunsheath with you guys."

My face lit up. Perfect. Caitlin coming solved our biggest problem.

"But please don't forget that it's my 25th birthday that weekend," she said.

"You got it, Caitlin. I promise we'll have a great time. We'll even plan a steak dinner on the trip."

The four of us started putting a plan together. We agreed to take two canoes: my dark green Royalex plastic Spirit II, which we would put on a plane and fly out with us after our vacation at Dunsheath Lake. For a second canoe, we'd buy a used one to leave there. The only problem with this plan is that canoes, even used, aren't cheap. And the cheapest canoes are usually the smallest and the heaviest. We ended up finding a fifteen-and-a-half-foot triple-layered dark green plastic canoe for sale on the internet. It was good enough to get us safely to Dunsheath Lake, but also cheap enough that we wouldn't feel bad about leaving it there to be used only a few times a year.

We picked our dates, laminated our topographical maps, made packing lists, planned a menu, and rented a satellite phone. The plan was to start as early in the morning as we could on the first day and make it halfway up Goldsand Lake. Warren had some work to do in Lynn Lake, so he'd meet us there with one canoe. Caitlin, Jason, and I would drive thirteen hours through the night to get there with the other canoe tied onto our vehicle's roof.

The three of us made a deal. On the drive from Winnipeg, I'd sleep in the backseat until Thompson. Then we'd switch spots. I'd drive the worst part of the road, the highway from Thompson to Lynn Lake, at the worst part of the day, from 2:30 a.m. to 6:00 a.m., and they would get to sleep in the back. Perfect.

In the dark of 3:30 a.m., I discovered that our vehicle didn't have a headphone jack or wireless capabilities. I rummaged around for a CD but found only one in the entire vehicle: the Lion King soundtrack, Broadway edition. I can appreciate the Circle of Life and Simba wanting to be a mighty king once or twice, but even Hakuna Matata begins to lose its meaning after a while. The endless repetition, along with the inability to sing out loud due to my sleeping passengers, was slowly making my eyelids close.

But I persevered. We made it to Lynn Lake safely. In hindsight, though, that musical oversight should have served as fair warning that things weren't going to go as planned on this canoe trip.

Warren had slept in a tent that night, and the weather had gone down to minus one degree Celsius. He woke up early, wanting hot coffee and bacon. We met him, tired and dishevelled, at the Route 391 Bar and Grill, also known as the Red Barn, for breakfast. They were closed.

We dropped the canoes at the launch point on Burge Lake and parked Warren's truck at the airbase for our flight back in a week. Hungry for breakfast, we headed back to town to buy some food from the grocery store. Luckily for us, the Red Barn was now open. With a long day of travelling ahead of us, we ordered four trucker's breakfasts.

After ordering, Caitlin asked, "What time do you open? We thought you'd be open early for breakfast."

"Oh. We open whenever the staff get here," the server replied.

So helpful, right? And that's why we started paddling ninety minutes later than we had originally planned.

WHO GOES CANOEING WITH THEIR MOTHER-IN-LAW?

Our trip started on Burge Lake. We posed for pictures on the dock, did pre-trip shots of brandy and set off. Caitlin was in the bow of the small canoe with Jason in the stern, while Warren paddled in the bow of my canoe. The only sounds on the lake were the loons calling each other and Caitlin sharing beloved childhood memories of summer camp on Burge Lake.

We paddled to the northeast corner of the lake and portaged over a beaver dam into the Stick River. The water was remarkably low, so we ended up getting out of our canoes and dragging them over mud, around rocks, and through stagnant water with depths that were only discovered when we stepped in them.

After a hundred metres of this we looked up and saw the rocks of a dry riverbed.

"Where's all the water?" Warren asked. We took out our maps and saw a black line that we thought would have been navigable.

"Looks like the water levels are low this year. I guess we're walking," Jason said.

So we pulled our canoes up onto the rocks for the time being and started walking where there should have been water. We were looking not only for the next body of water to paddle in, but also for an easy route to get there.

We climbed over rocks. We ducked under trees. We hopped over puddles. Deadfall was everywhere. We dreaded the thought of carrying our gear through this. As we walked further and further away from our canoes, we were well aware that every step now would be a portage step later.

"I found the lake!" Jason called from up ahead. We scrambled up behind him. There was Stick Lake, still exhibiting the calm waters of the morning.

We decided to spread out and see if there was a better route to portage our gear other than the dry riverbed. Sure enough, we found a snowmobile trail about fifty metres to the east, so we clambered back for our stuff. Warren led the way, earning the nickname of "Russian Bear" by picking up the fallen trees in our way and throwing them aside like toothpicks. Warren is actually of Ukrainian descent, so how he ended up as the "Russian Bear" is a bit of a mystery to all of us.

We hadn't planned on portaging the first day, as evidenced by the bag containing several litres of homemade wine in a cardboard banana box, so we wisely chose to make the portage in two trips.

Stick Lake was calm, and we enjoyed chasing a flock of ducks for several minutes. While they flapped away from us quickly, they always landed just ahead of us in our direction of travel, only to flap away once again.

We paddled around a bend in the lake and shook our heads at what we saw ahead of us. There was nothing but tall, green reeds! We paddled straight into them, but our heavy-laden canoes quickly bottomed out. Unable to move, we got out and pushed. Our laughter at the absurdity of walking outside our canoes on the reeds was soon diminished by another set of un-portage-able rocks.

Stupid Stick River, with no flowing water.

Warren and I trekked up the dry riverbed to Goldsand Lake, found the snowmobile trail, and walked back to Stick Lake. We were happy to see that our canoes were only fifteen metres from the path. We portaged again.

We stopped for lunch at the southeast corner of Goldsand Lake. The two unexpected portages had left us four hours behind schedule, but, anticipating the late sunsets of the north, we powered on. I wish I could make the next five hours seem more exciting than they were, but really, this is what we did on Goldsand Lake: We paddled north for one kilometre, did a 90-degree turn, and paddled straight east for five kilometres. We did another 90-degree turn and paddled straight north for another seven kilometres. Our routine was to paddle for fifty minutes and then rest for ten. We repeated this pattern over and over again, with only a few islands to break up the monotony. Surrounded by forests of black spruce, jack pine, and tamarack trees with hills rising in the distance, the scenery was initially appealing. However, after the second hour, everything started to look the same. To keep ourselves engaged we changed how many strokes we did on each side before switching. I know, right? Pretty exciting stuff.

It was nearing supper time when we reached the narrows between the two large bodies of water on Goldsand Lake. There are small, natural beaches throughout the lakes in Northern Manitoba, and we had our eyes peeled for one where we could spend the night. We found one three kilometres further up the eastern shore of the western half of the lake, sheltered behind two islands. It was small—only there because of the low water—but it was manageable. A few metres into the bush we found a flat spot for our tent and settled in for the night.

As promised, I made steaks for supper that night! But, when serving them, I accidentally dropped them in the sand. A quick rinse in the lake didn't remove the sand very well, but it certainly removed the heat! We pretended not to mind eating wet, cold, and sandy steaks, and spent the rest of the evening spitting sand out of our mouths. Happy Birthday Caitlin!

After supper we met to plan how the next two days would look. We pulled out the maps and laid them on the overturned canoes. I started measuring distances and assessed our current situation.

"Okay. We paddled and portaged twenty-five kilometres today. That took about ten hours. Good job everyone. We worked hard, but are quite tired now," I said.

"We'll sleep well tonight, but maybe that'll be cancelled out by how sore we're going to be in the morning," Warren noted.

"The wind was calm today. If we get bad weather, that'll slow us down," Jason said.

"Let's see how far we have to go in the next two days," I said, looking at our map. I counted the squares on the topographical map as best I could. "We have about fifty-five kilometres to go to get to Dunsheath. We have five known portages to get from Goldsand Lake to Wells Lake, and then an unknown number of portages once we're in the Hughes River system—except for these two waterfalls that we know of near Dunsheath Lake. Plus, low water."

Silence.

"Those are two long days ahead of us," Caitlin said.

"When we get to Dunsheath, at least we'll have food and shelter waiting for us," said Warren.

"Can you imagine what our families would do that day? They'll probably be sitting in the boats all day at the river that flows into the lake, waiting for us. What will they think if we show up late in the evening?" Jason said.

"Or what happens if we don't make it in two days, but three? If we show up one day late, they'll be worried sick about us," I said. "I don't think Ashley would ever forgive me."

"We have the satellite phone. But once they leave Thompson they won't have any cell reception. If we don't call them before they leave Thompson tomorrow morning we won't be able to notify them if our trip takes an extra day," Warren said.

"That would be bad," Jason looked out over the lake. "Very bad."

Caitlin looked out across the lake and let out a heavy sigh.

"I see three options," I said. "We either paddle hard for the next two days to Dunsheath as planned, we paddle hard back to Lynn Lake tomorrow, or we call for a floatplane to come pick us up. But we have to decide tonight so we can let our families know what's going on."

We silently weighed our options.

"I'm out. I just can't do it," Caitlin choked out, tears rolling down her cheeks. "I can't do this for two more days, or longer days, or any more days. I'm done."

We waited until her tears had dried up.

"Is there anything we can do to convince you otherwise?" I asked, trying to figure out the best way to respond.

"No. I'm sorry. I'm done."

The four of us looked at each other, unsure of what to say next. Our eyes sensed motion in the forest. A grey jay was darting between trees, watching us.

"Okay," I said. "Then we're done. It's alright. We understand. We'll call your dad to send a plane for us tomorrow."

I turned on the satellite phone and called Doug. After explaining that we were okay, but needed to be picked up, I gave him our coordinates. He said that he'd call the airbase in the morning and have someone come pick us up as soon as possible.

Exhausted from the hard day's work, and only getting a few hours of sleep the night before, I replayed the day in my mind, over and over again. What could we have done differently? Did we plan poorly? Were we overly ambitious? Or was it just because of the late start and cooks that slept in? Thankfully, it was only a few minutes until sleep won out.

We woke up in the morning with a bit more jump in our step. Sure, we were getting picked up, but we still had a week of vacation at Dunsheath Lake ahead of us. All was not lost. We had breakfast and packed up our tent. We checked the phone, and Doug had sent us a text that a plane was coming to get us in the morning. We had no idea exactly what time that plane would come, so we kept ourselves busy by fishing and reading.

I noticed that both our canoes were dark green, and our tent, beige. Those weren't the colours to choose if we wanted to be spotted from the sky. I climbed a small rock hill just to our left and laid out our bright red and orange life jackets for the pilot to see.

An hour later, we heard a distant rumble coming from the southeast. We couldn't see past the trees behind us, but we knew it was the sound of a floatplane. Caitlin, Jason, and Warren stacked our stuff in a neat pile on the beach. The engine noise was getting louder and louder. I ran up the hill to collect the life jackets, and as I got to the top, the plane buzzed right over me, waggling its wings.

I cheered. The three on the beach cheered. The plane circled us. We were rescued!

The plane circled three times, turned around, and pointed itself back to where it came from. It flew back over us one more time, and then was out of sight again.

"The pilot must be looking for a safe landing place," Warren said. "They'll circle for a bit to check for rocks, and then land here by the beach."

The noise of the engine faded. We strained our ears to hear the plane turn around for us, but soon it was silent again.

"What's going on?" I asked. "Where did it go?"

"Maybe this was just a scouting trip and it'll come by later to pick us up?" Caitlin wondered.

"I saw the wings waggle! Did you guys see that?" I asked.

"Yup."

"Well, the waggle means the pilot saw us, and it circled us, so we're good. It'll come back for us. Maybe they just have to do some other runs first."

We unpacked our fishing rods and books and resumed waiting.
We waited.
And waited.
And waited some more.
Caitlin quietly sang happy birthday to herself.
We had lunch.

WHO GOES CANOEING WITH THEIR MOTHER-IN-LAW?

We waited some more.

We were quite confused as to what was going on.

"Look!" Warren jumped up, pointing to the sky. "A floatplane!"

We saw another floatplane, far off on the western horizon, just above the tree line. What was our plane doing way out there, when we were over here?

The plane was flying north, away from us, and then it disappeared. Strange, we thought.

We waited some more.

The same plane reappeared, now flying straight south. Again, not flying anywhere near us. We figured that must be a different plane, picking up and dropping off guests from a fishing lodge.

By now it was mid-afternoon. It had been four hours since the plane first flew over us. Something wasn't right. It had easily had enough time to go to the airbase in Lynn Lake, pick up our family, drop them off at Dunsheath, and come back to get us. The flight to Dunsheath was only twelve minutes long, and, as the crow flies, we were only twenty kilometres away.

The fishing lodge plane appeared again in the distance, heading north again. So frustrating!

We put the canoes out again and fished. We took naps. We played tic tac toe in the sand.

The fishing lodge plane headed south again, and we gave up any sliver of hope that that plane was coming to get us.

Warren started doing bicep curls with rocks on the beach. He marked the first curl, saying, "Bored." More bicep curls. "Bored. Bored. Bored."

He stopped. He tilted his head up. The Russian Bear apparently has ears like a fox.

We looked up, too, hearing what Warren heard first.

The engine of a plane from the southeast. Getting closer. Just like the morning.

91

"Well, it's about time," I said.

The plane buzzed us, identical to the morning. It flew over us and turned 180 degrees again, identical to the morning. It flew back over us again, out of sight, identical to the morning.

The plane did not come back. Again. Identical to the morning. We listened to the engine noise slowly disappear.

"Oh, come on. Not again!" Caitlin yelled, waving her hands in the air. "Come back! We're here!" Her arms fell back to her sides. "Worst birthday ever," she muttered.

"WHAT IS GOING ON WITH THAT PLANE?!?" I yelled to no one in particular. "WHY DOES IT KEEP COMING AND THEN LEAVING WITHOUT US?"

An hour later, we got our answer.

We heard the engine of a plane from the southeast. Getting closer. Just like the previous two times.

The plane buzzed us, identical to the morning and afternoon. It flew over us and turned 180 degrees again, identical to the morning and afternoon. It flew back over us again, out of sight, identical to the morning and afternoon.

We waited as the engine sound grew quieter. We were losing hope of ever being picked up.

But the rumble of the engine didn't disappear entirely this time. It was actually getting louder again! It was coming back for us!

We cheered again and hugged each other on the beach. This time we were rescued for sure!

The plane circled a few times looking for rocks, as we thought it would do in the morning. But this time it landed and taxied to the beach. When the door opened, the pilot stuck his head out and put a Fig Newton in his mouth. "Want a ride?"

We loaded up our gear in the back of the plane and the pilot said he'd come back for the canoes later in the week. We stashed them in the forest, tightly wedged between some trees.

Before takeoff, I asked, "You flew right over us this morning. Did you have a lot of flights today that you had to get to first?"

"Nope. I didn't see you at all. It's really hard to spot things from the sky." He pulled another Fig Newton out of the package and put it to his lips.

"But you circled us three times!"

"As I said, it's hard to see people from the sky. It's best if you put your boats in the water. That's the easiest way for pilots to see you down there," he said, now chewing his Fig Newton.

"Oh. But didn't you waggle your wings at us?"

"Nope. At least, not intentionally."

"Huh," I said. I was certain I had seen a waggle, and wondered how he had missed us with his three circles.

We landed on Dunsheath Lake a few minutes later. The low water meant he couldn't get the plane too close to shore. We got out and walked, the water up to our waists.

I will never forget seeing Arianna jump up and down on the shore in excitement, yelling "Daddy! Daddy! Daddy! Daddy's here!" I ran through the water as fast I could, not caring about how wet I was, and wrapped her almost three-year-old body in my arms. Her face nuzzled deep into my neck.

"Daddy!"

"I'm here, Arianna. I'm here, and I love you."

Zach was just a five-month-old baby content in Ashley's arms, but he got a big hug too.

I turned to Ashley and held her for a long time.

When the time for words reappeared, I asked what had all happened that day.

She rolled her eyes. "Dad sent the plane out early in the morning, and the pilot came back to the airbase in Lynn Lake saying he couldn't find you. We were scheduled for two flights with all our people and gear. Joanne and I were on the second flight and it detoured over Goldsand Lake to go and look for you. As soon as we were near the spot you told us on the map, I saw you immediately. The pilot didn't even notify me that we were close. I really don't know how he missed you the first time."

I grinned. "You're amazing Ashley," I said. "Thanks for rescuing us." She smiled back.

After we unloaded the plane and brought everything up to the cabin, we sat by the fire and took a few deep breaths. The mix of emotions was confusing: relief, gratitude, frustration, joy, defeat, acceptance.

I wrote down the details of our canoe trip in the Dunsheath Lake journal, spinning the angle of the story just a bit.

"Part of me was discouraged by the seeming failure, our plans and dreams left unfulfilled. This is the first canoe trip I have failed to finish. But, it wasn't a failure because we couldn't do it. It was a failure because we chose to get picked up so we could get to Dunsheath Lake in time to be with our families. Any feelings of failure were quickly replaced with the anticipation of reunion and being at Dunsheath. Our final thoughts about our canoe trip to Dunsheath Lake? Simply postponed."

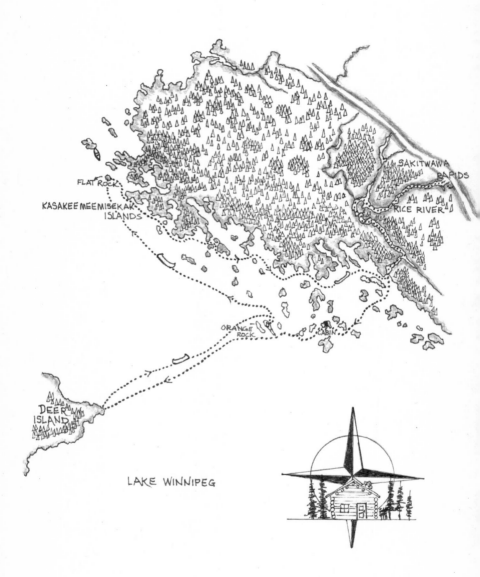

– 9 –

Somebody Stole Our Canoe

June 2016

Ashley was on sick leave while pregnant with Milo, and I had a week of vacation to use up before he was born. I spent four of those days with Ashley, Arianna, Zach, and in-utero Milo. I snuck away for the next three days to go canoeing with my friend and Ultimate Frisbee teammate Randy.

With three full days of mid-summer sunlight ahead of us, we had many options of where to go. I wondered about the Mantario Wilderness Zone in the Whiteshell Provincial Park, and Randy offered to take me to the Experimental Lakes Area east of Kenora. In the end, we decided to go somewhere neither of us had been before.

Several years earlier I had read the book *A Daytripper's Guide to Manitoba* (Great Plains Publications) by Bartley Kives, a journalist, author, and avid paddler. I remembered that he had included a few sections on paddling, so I leafed through it again. One little paragraph caught my attention.

> *Two kilometres west of English Brook, a winter road heads north from 304 to the Bloodvein River, crossing the Rice River thirty km up. Canoeists and kayakers park here, paddle about five km down to Lake Winnipeg, and explore a gorgeous archipelago called the Kasakeemeemisekak Islands, where amazing numbers of white pelicans, cormorants, and bald eagles gather during the summer. The islands make for a fantastic day trip, as long as you're prepared to deal with rough water on Lake Winnipeg and the navigational challenges of finding your way in this maze. Bring a GPS, just in case.*

Further research confirmed the presence of river otters, common terns, moose, and bear.

I pitched this idea to Randy. His response was an immediate yes.

I drove from Steinbach to the West End of Winnipeg through the morning rush hour. You'd be amazed at the funny looks you get while driving down Broadway with a 17-foot canoe sitting on top of a Honda Fit. I picked up Randy and we made our way out of the city, stopping only in Powerview-Pine Falls as we drove up Highway 304 past Manigotagan.

When we turned left onto the Bloodvein Road at English Brook we grinned with giddiness. I had only driven on this road once, in winter, and in the opposite direction, so every turn of the car this time brought a sense of anticipation and wonder. Over every bridge we checked both directions to look for waterfalls and future paddling destinations. We watched for moose, deer, bear, and oncoming traffic, in that order.

At kilometre thirty, we saw what we were looking for. Alongside a bridge was a brown sign with the words "Rice River" written in white. We pulled over on the south side of the bridge. Upriver the water was flat and slow-moving. Under the bridge and downstream, the Sakitwawa Rapids roared at us. Two eagles took off from their pine branch perches and flew towards Lake Winnipeg.

Many of the rivers on the east side of Lake Winnipeg are known as "pool and drop" rivers. To paddlers, a "pool and drop" river means that a river can be paddled easily with a gentle current pulling you along, and then every so often a quick elevation change results in either running the rapids or portaging around a waterfall.

The road was plainly built over one of the "drops" in the Rice River. We saw an opening in the bush ahead. Closer inspection revealed a clearing to park our car, and a short portage trail to the Rice River. A calm and bubbly "pool" met us at the bottom of the waterfall. We parked our car, carried our gear down to the river, and were off.

One of the most exhilarating things you can experience while canoeing is the exploration of new canoe routes. Every bend brings adventure, every fork in the river makes you pull out the map, every tree is new, every cliff is worth paddling under, every unmarked beaver dam is a challenge, every hill is scanned, and every stroke of the paddle brings you further into the unknown.

Bald eagles were everywhere. Some were perched royally, sentinel-like, on the lookout for fish swimming near the surface. Others were juveniles, still spotted and brown in colour, yet just as large as the adults. We tried sneaking up on some to snap their pictures. But bald eagles are skittish birds, and most of them heard us coming long before we could get close enough.

We stopped on a beaver lodge to eat our lunch of naan bread and hummus. We wondered if there were any beavers inside, hiding deep in their den away from the unfamiliar stomping above. Another eagle flew by, the swooshing of its wings breaking the silence. We could not stop smiling.

We paddled on, idly dropping our fishing lines in the water in hopes of catching something. Surely the eagles were there for a reason, but our spoons and spinner baits kept coming back empty.

We were in no rush, so we slowed down to take it all in. I even set up my Therm-a-Rest sleeping pad as a chair, with back support and all, and leaned back to examine the scenery, counting the eagles in the sky. The river beckoned us on, first under a cliff on our left, then beside a gentle hill lined with pine trees on our right, then by a rock in the middle. Just past the rock the river opened up. Our jaws dropped. The blue-green waters of Lake Winnipeg were broken up by islands rising out of the lake. They were all similar, with rocky shores giving way to thick mini-forests on top. Kasakeemeemisekak means *Many Islands* in Cree.

Eagles flew by again, but this time they were accompanied by pelicans, seagulls, double-crested cormorants, and common terns.

"Which way should we go?"

Shoulders shrugged.

"South it is!"

Lake Winnipeg, the tenth largest freshwater lake in the world, is an inland ocean. It stretches over 400 kilometres north/south, and up to 100 kilometre east/west. The east side of the lake is generally Canadian Shield, with millions of acres of boreal forest growing on granite bedrock. The west shore is called the Manitoba Lowlands, where limestone cliffs, swamps, willows, and aspens dot the shoreline.

The Kasakeemeemisekak Islands are in a bit of a transition zone between the two, where parts of them are clearly Canadian Shield, other parts are clearly Manitoba Lowlands, and some are a combination of the two. It's truly a unique and fascinating place.

Paddling to the Kasakeemeemisekak Islands on a calm day was simply beautiful. We paddled past the first island, with nothing on our early afternoon agenda other than to explore and eventually find a place to sleep. We found a small beach on the second island

and scrambled out to explore. As we approached the third island, we were greeted with a sharp hissing sound from the forest.

"What is that?" I asked.

"I have no idea," Randy answered.

We peered into the underbrush of the forest on the island. We heard a small animal running back and forth in the forest, stopping at each end of its path to hiss at us. Unable to see it, we couldn't figure out what it was.

"Maybe it's a fisher? A martin?" I suggested.

"Didn't we read somewhere that there were river otters here? Maybe it's one of those," Randy said.

In an attempt to see what animal was really bothered by our presence, we maneuvered the canoe back and forth to change our vantage point. Still nothing but hissing. But in between some rocks near the shore, we heard some whimpering. Randy got out of the canoe. He saw a small ball of fur cowering between the rocks. A fierce hiss from the forest followed him. He jumped back into the boat.

"Well, at least we know why it's mad at us. Whatever it is, that's her baby, and she doesn't want us anywhere near it. We should probably leave it alone."

We paddled on, slowly, hoping to get a glimpse of the angry parent coming to console its scared child; but alas, nothing.

Randy pointed. "What's that big orange thing in the distance? You can see it, just past the island in front of us. What is that?"

I followed his finger, and sure enough, sticking out of the water was a rock bigger than any of the islands, so high that we could see it above the rest. It was bright orange.

"Let's go see!"

We did our best to paddle in a straight line towards the big orange rock, but our path was interrupted by other islands calling for our attention. And, on one of them, we saw a cabin.

"Let's go check it out!"

We landed at a small beach, pulled our boat up, and walked on freshly cut grass dotted with Canada goose feces. We peered in the windows of the cabin. Nobody seemed to be home. We tried the door. It wasn't locked. We walked in.

We were greeted by a rustic, two-bedroom cabin. One room was filled with bunk beds and a small kitchen, and the other was filled with a single bed and a wood stove. The interior walls and studs were covered in graffiti, which, upon closer inspection, revealed phrases like, "Thanks for the great weekend Harvey!" and "Our grade 8 class was here at Harvey Bushie's cabin, 2002."

Looks like we found Harvey Bushie's cabin. But who was Harvey Bushie?

Feeling a bit like trespassers, we finished exploring the cabin. Outside, we found a path that led to the other side of the island. Through the trees, we could see, and hear, hundreds of pelicans and seagulls gathering at a peninsula at the end of the island.

I had an idea.

"Randy, you wait here with the camera. I'll run down the path and bark like a dog. I'll scare all the birds into the air. Then you can take some great pictures of all the pelicans taking off. Okay?"

Randy agreed and unslung the camera.

I crept down the path, my eyes focused only on the birds. When the first ones heard my footsteps and turned their heads towards me, I ran out.

"WOOF! WOOF! RUFF! RUFF! WOOF!" I barked as loud as I could, running down the path. Some of the birds started to fly.

"RRRRUUUFFFFFFFF!!!!" I bellowed out.

And then I saw them. On the edge of the peninsula—two men filleting pickerel on a dock with a commercial fishing boat attached. They turned their heads at the commotion. I turned on a dime and ran back down the path, right past Randy.

I shouted over my shoulder without slowing down, "Shoot, Randy, there are people here!"

We ran back to our canoe, and as we got there we turned to look back at the path. Two men were walking towards us, looking puzzled. They were First Nations, about 40 years old, both with rubber overalls on, one with a toque and one with a ball cap.

Randy and I took a few breaths, gathered our composure, and walked towards them, smiles on our faces.

"Hello! How are you guys?" we called.

"Good, good." They answered. We shook hands. They looked at our boat. The man with the ball cap asked, "Did you guys canoe here?"

We glanced at our canoe, and back to them. "Yeah. Yeah we did. You don't see many canoes out here, do you?"

"No. No we don't."

The man with the toque jumped into the conversation. "Were you guys barking like dogs just now?"

Randy smiled. "Ummm…yeah! That was actually him," pointing to me.

Both men looked at me. "Yeah, so…we just wanted to take a picture of the birds flying. You know. We're from the city, and we don't see pelicans all that often. I didn't know you were there. Sorry about that."

They smiled, probably thinking that these two white guys were crazy, paddling out here on Lake Winnipeg and barking at pelicans. In hindsight, that wasn't an inaccurate description.

"Who's Harvey Bushie?" Randy asked.

"Our boss. The guy who owns this cabin." It turns out that Harvey Bushie runs a commercial fishing outfit out of Hollow Water First Nation. These guys set fishing nets, and the next day they come back and gut the fish. After keeping what they need for their families, they sell the rest. This island was his home base for the area.

"It's a pretty nice place you guys got here," I said.

"Yeah. Where are you guys paddling to?"

"We don't know. Know any good spots?"

"Yeah, right here. You even have a cabin you can stay in if it rains!"

It was our turn to smile.

"Thanks for the offer, but we're going to paddle to an island a bit more out of the way," Randy said. "But thanks anyways!"

We shook hands again and went separate ways. They went back to filleting their fish, and we went back to paddling. Out of earshot, we started laughing.

"Why the heck did you think barking like a dog was necessary, Kyle? You could have just yelled anything, you know."

"Shut up, Randy."

Once more, we pointed ourselves towards the orange rock—but this time we had fun aiming our canoe through as many flocks of pelicans as we could. We could hear the slap of their feet running on the water, their big wings flapping, eventually lifting them airborne. Common terns flew by us, looking for fish. When they saw one, they'd flap their wings really hard to stop their forward momentum, look down, hover in place, and then dive beneath the water to catch their prey. When they came up with a small silvery fish in their mouths, they'd simply float on the water and eat their hard-earned meal.

We finally made it to the big orange rock. We dragged our canoe up a rocky slope, got our cameras, and climbed to the top. The rock was a big hump sticking out of the lake, covered by orange lichen. We took panoramic pictures of our surroundings, picturesque islands on one side, open water on the other.

Randy had taken his pictures with his phone, and noticed that he had full bars of cell reception. It must have been from the cell towers in Hecla Provincial Park. Even though we had come from the middle of nowhere on the east side of Lake Winnipeg, we were only twenty kilometres from Gull Harbour in the Hecla-Grindstone Provincial Park. The cell reception seemed to travel quite well over water.

We looked west, towards a distant shore across the open water.

"What's the white stuff out there?" I asked. "Is that a beach?"

Randy pulled out the map. "That's Deer Island. It's just over three kilometres away."

"That's for sure a beach, Randy. Do you think we can make it?"

We had a healthy fear of paddling on the open water. Lake Winnipeg is like an ocean, and a little wind can whip up big waves in a short time.

I pulled out my phone and checked my weather app.

"I think we're good for today. Minimal wind."

We looked at each other, silently assessing how much risk the other would take to get to that beach. We grinned.

"We can make it."

"Let's do it."

We left some of our gear on the orange rock, including our cooler, tent, and bags, taking only our water, fishing gear, navigation tools, life jackets, water bottles, and my Therm-a-Rest chair. We pushed off into the open water.

Luckily, it was a calm day. With the bow pointed towards the beach and a slight tailwind behind us, we made it to Deer Island in less than an hour.

We couldn't believe what we had stumbled upon. Kilometres of a beach made of fine, white sand, clear of debris, and it was empty! We were hot from both the paddle and the sun, so we waded out 25 metres to the deeper water, walking on nothing but sand the entire way.

I was astonished. "This is like a tropical beach. But it's in Manitoba!" I said.

"And we have it all to ourselves!" Randy yelled to nobody in particular.

After a quick dip to cool down, it was time to fish. Randy hadn't done much fishing in his life, so he was looking to me for guidance. However, most of my experience was in Northern Manitoba where catching a fish every cast was the norm. I did know enough about fishing, however, to be skeptical about our chances of catching fish in the shallow waters of this beach. But with nothing else to do, we gave it a shot.

We were up to our knees in the water, casting simple jig heads with a plastic worm, quite content with the simple rhythm of casting out and reeling in. Off to our left, a flock of common terns drew our attention, all hovering in the same area. All of a sudden one dove down, and another, and then another! Most of them were coming up with a fish in their beaks.

"Hey, Randy! I bet there's a school of bait fish there! Kind of cool, eh?" I said.

We watched the birds feast for a minute or two, hypnotized by the pattern of hovering, diving, eating, and then taking off again. As we watched the birds, we started to notice some bigger splashes in the area, too big to be from the bait fish, but not at all related to the birds.

I wondered to myself what would happen if I cast my hook into the chaos of the birds and the bait fish. I cast out, and immediately felt a tug on my line. I jerked my rod, setting the hook. A silvery fish breached the water, flapping wildly before slapping against the surface tension of the lake.

"Randy! There are fish chasing the bait fish from the bottom! I bet they've got those fish swimming in a tight ball. Like dolphins on Planet Earth! Get your line in there!"

He did, and instantly he had a fish on his line too. We started laughing at the absurdity of our double hit. I had reeled my fish in at this point and saw that it was a white bass. I had never caught one before! I let it go and threw my line back into the chaos. Another one!

Randy had released his fish, and he too shouted for joy that he had another hit. We reeled these in with huge grins on our faces. I realized that we had no idea what the Master Angler size for white bass was. We walked onto the sand with the fish, put them on a paddle and took a picture to measure when we got back home. Walking back into the shallows, we released the fish.

We waded back up to our knees, our eyes following the flock of terns. They had stopped diving. We threw our lines underneath them and came up empty. After a few casts, Randy suggested that we wait a bit and watch the birds.

They hovered over a different part of the beach, about twenty metres to our right. And then one dove. And another. And another. A white bass jumped up from the water.

"There!" Randy yelled. "Let's go!"

We ran to where the water was ankle deep and sprinted in the direction of the terns. With the water splashing around us as we waded deeper, we cast our lines. Boom. Boom. Fish. Fish. And mine was HUGE!

We couldn't believe our luck! We were on an island all by ourselves, on a beautiful beach, running around in the water with our fishing rods catching fish with almost every cast while stalking some birds.

After landing fifteen fish in thirty minutes, both the flock of terns and the bites on our fishing line thinned out. The feeding frenzy was over.

We had spent more than two hours on the Deer Island beach and it was now late afternoon. We kind of regretted leaving most of our camping gear back on the big orange rock, as camping on the beach seemed quite alluring. But with our overnight gear lying across the open water, we packed up and paddled back.

The slight tailwind was now a slight headwind, though very manageable, and we made it back in an hour. The evening was upon us. Since we had been up since sunrise we were getting pretty tired. We found a camp spot one island inland from the orange rock. Here we set up our tent and made supper. That evening, Randy taught me a wonderful habit that I try to do now on every canoe trip: I pour myself a glass of wine, set up my chair facing west, and watch the sunset in silence.

Because we were canoeing close to the summer solstice, the sun took a long time to set. A really long time. Given that I had young children at home and Ashley and I hadn't slept through the night in months, I started to nod off as soon as the glorious oranges and reds of the sun started to fade.

"I'm going to bed, Randy. Good night."

"Good night, Kyle."

And that was the last sound I heard until morning, when Randy's voice called to me through the tent door. "Breakfast is ready, Kyle."

Randy is a good friend.

We decided to paddle northwest that day, as the map showed more islands in that direction. We slowly paddled to and fro, with nothing compelling us other than eventually finding a flat piece of land to set up our tents.

We found ourselves in a bay and spent an hour trying to catch some carp that kept surfacing. We think they were breeding, as there would be a small commotion of two fish circling each other, and then they'd float near the surface. Being bottom feeders, our tackle was useless, and we couldn't quite get one of our hooks to stick into them. We didn't have a net to grab one either. We did get close enough to one to hit it with a paddle, but that just scared it to the bottom.

Surrounded by eagles, pelicans, and cormorants, we paddled along past islands that were growing smaller with fewer trees.

Given the uniqueness of the Canadian Shield ecosystem meeting the prairie ecosystem, the trees that were growing had transitioned from pine to birch, poplar, and oak.

We came across a long, narrow, flat island with a large, solitary oak tree on it. That oak tree must have been unbelievably hardy to survive the cold, windswept winters on the lake, as well as the car-size boulders of ice that the lake chucks around every spring.

Under the tree was a bit of a low flat spot, a perfect site to pull up our canoe and unfurl our camp kitchen. The rest of the island was about one metre higher than where we prepped our food, all of it flat and covered with moss and lichen and low-lying shrubs. Best of all, we had an unobstructed view of where the sun would set that evening. Again, we couldn't believe our luck.

I made supper while Randy set up our tent. I filled some tortillas with rice, cheese, and refried beans, wrapped them in tin foil and put them by the hot coals of our fire. I set up the wine box on the bow of our canoe, poured two glasses, and went to check on Randy. He had just finished stretching out the fly of the tent. Since we were camping on a thin layer of moss on top of rock, he was pretty pumped to have found some rocks to place on the corners of our tent.

"Randy, where did you get all those rocks from?"

"Oh, just over there. It looks like they're leftover from a fire ring years ago."

I looked around. I saw the circle of rocks he was referring to, three metres in diameter. The circle was now missing three or four rocks, but there was no sign that a fire had been lit there before. No ash, no charcoal, and no scarring on the rocks.

"Uh oh, Randy. I think this might be a petroform." Petroforms are ancient rock arrangements, usually in the shape of an animal, and are considered sacred to First Nations.

His face dropped. "Oh no!" We looked closer.

"Maybe those are the legs, and there's the head, and it's a turtle? And, there is an oak tree here, and I've heard that oaks grow where First Nations historically camped."

We were mortified. Embarrassed. And so remorseful. Words cannot express our regret.

We quickly put the rocks back how Randy had found them, even doing our best to get the correct side up by matching the lichen patterns.

We stepped back and assessed.

"Yup. Turtle."

"If anybody is listening, we are so sorry. That was an honest mistake, and one that we won't make again."

We went down to the kitchen site under the oak tree and enjoyed our supper, a bit sheepish, but thankful that we were able to replace the rocks. After we had cleaned up, Randy suggested that we take our wine glasses, kick off our shoes, and tour the island before setting up our chairs to watch the sunset. I agreed.

The island certainly was unique, being so flat and barren. We started walking counterclockwise, looking out towards Deer Island, and then slowly turned south. The south wind was dying, but still gently pushing the waves up against the island. We stopped for pictures often, sipping our wine and marveling that, minus the fishermen, we had this entire archipelago all to ourselves.

When we were about three quarters of the way around the island we turned back north to our campsite for dessert. We could see our tent, but not our canoe or kitchen as they were hidden behind the oak tree on the lower rock shelf. In addition to our tent, though, we also saw a man, backlit by the sun, solo-paddling a canoe near our island.

Who was this guy? Where did he come from? We were on an island in the middle of Lake Winnipeg, by ourselves, a three-hour paddle away from our car, and we hadn't seen any other canoeists all day. This was strange.

We continued to watch him.

He wasn't moving. He just stared back, slowly drifting away from our campsite with the wind.

We waved. Nothing.

Something wasn't right. Was this guy coming to steal our canoe? But how did he get to our island without his own boat? We hadn't heard any motors. Was he trying to steal our camping gear? Where would he take it to? Surely he had seen us. Other than the oak tree, we were the only things taller than shrubs on this island!

We looked at each other, and agreed that we needed to get back to our campsite to see what this guy was up to.

We started running.

"OW! OW!"

"ARGH!"

Searing pain ripped through our feet. We looked down.

Cacti. Little cacti, that stick into your skin. The needles were so sharp that we couldn't use our fingers to take them out of our feet. The prickly pear cactus is native to Manitoba, but we didn't know it was native to the Kasakeemeemisekak Islands. Now we did, as the cacti were embedded in the soles of our feet.

After yelling in pain, we pulled out our multi-tools and started removing the cacti from our skin. While Randy was looking for a cacti-free place to take his next step, I found a path back to camp.

When I got there, my heart stopped.

"RANDY!!!!!! THAT'S OUR CANOE!!!!!"

It wasn't a man paddling! It was my Therm-a-Rest sleeping-pad-turned-chair that looked like a person! It was backlit it by the sun, so all we saw was a silhouette! A shift in the wind probably raised the water level a few inches, and our canoe had slid into the water. Now it was drifting away into the sunset.

Randy found his way back to the campsite. We needed that canoe. We needed to go get it, now!

Lifejackets! We looked around. Mine was lying on the rocks, but Randy had left his in the canoe. One would have to do.

Paddles! Again, we looked around, but couldn't see them. They must be in the boat.

The canoe continued to drift north as we kicked off our clothes. I put on my life jacket and promised to share with Randy if he got tired, and off we went.

I ran full speed down slippery rocks into the water and swam to the next island, straight east of us. After kicking a few underwater rocks, Randy suggested to me that we didn't need any broken toes or sprained ankles. I copied his lead and slowed my water entries and exits considerably. We waded to the next little island, this one a little more north. When we had gotten as close to the canoe as we could via island hopping, we started swimming.

And swimming.

And swimming.

We looked up from the water, but could no longer see the canoe anymore. Where did it go? Did it tip? Did it go behind an island? Did the wind take it to the open water?

While swimming, I squinted. "I see it! There it is! Against that pile of rocks. Let's go!"

And so we swam some more.

About halfway to where I thought the canoe was, I realized that what I had seen was just a large rock. While I hadn't forgotten to take off my glasses in the excitement, I had forgotten that I am nearsighted. Lucky for me, Randy saw the actual canoe close to where I had thought it was. In desperation, we kept on swimming.

As we got closer we saw that the wind was bumping the canoe against the rocks and realized that we were going to be okay. We slowed our pace. We took deeper breaths. When our feet touched rock and we were able to stand, we let out cries of sheer joy and elation! We even found the box of wine floating beside the canoe. The wind had pushed it along with the boat! We poured that wine into our mouths with such reckless enthusiasm that it dripped down the sides of our faces. We made it! We got the canoe back!

We jumped into the canoe, found our paddles there as expected, and paddled back to our island. As we paddled back, we wondered if this was our punishment for disturbing the petroform. If it was, we'd gladly accept the lesson. Back at our campsite, we hauled the canoe far out of the water and tied it to the oak tree with two different ropes.

Hearts still pounding, we talked about what happened. What would we have done if our canoe was too far to swim to? Or if we had woken up and our canoe was gone? How would Ashley and Noelle have reacted if we'd had to call them and tell them to find the phone number for someone named Harvey Bushie from Hollow Water First Nation and ask him to come and get us?

We finished our wine and watched the sun set, ending the day with a large bonfire made from years of driftwood that had accumulated on the island. We went to bed with adrenalin still pumping through our veins.

The next day we broke camp early to head home. It was a calm, grey-sky day, with a few moments of drizzle. The glass-like conditions made paddling on Lake Winnipeg peaceful. We paddled for two hours to the river mouth, and then an hour upstream to where we had put in our canoe 48 hours earlier. We met a moose along the way. It blinked at us slowly before gently turning back into the bush.

The landing spot was at the bottom of the Sakitwawa Rapids that ran under the road. I had caught pickerel at the bottom of waterfalls before, and told Randy we should try fishing here. I checked my watch and set a limit of ten minutes.

We were jigging on the bottom of the river when Randy's rod bent. Like, really bent. His mouth formed an O, his eyebrows shot up, and his eyes grew wide.

"It's a big one!" he said.

I've caught hundreds of pickerel in my life, and I knew there was no way that even a ten pound pickerel should bend his rod like that. But I also knew it wasn't a northern pike, as they tend to hit the lure hard, drag your line out, and then act like sunken logs when you reel them in. At least until their next burst of energy.

"Randy, I don't know what kind of fish you have on there. Bring it up slowly. We're in a canoe and we don't have a net, but if you bring it to me, I'll do my best to grab it."

He fought the fish for several minutes. Every time we got a glimpse of it a few feet under the water, it dove back down to the bottom.

"What do you have on your line, Randy? Because I have no idea."

"Well, we're about to find out, because here it comes!"

He lifted his rod up. The fish didn't dive. I saw whiskers come out of the water.

"Holy crap! It's a catfish! Awesome! But I've never caught a catfish before. I don't know what to do! Do they have teeth?"

"I have no idea, Kyle. You're the one who fishes! What should I do?"

"Well, I think I can stick my hand in its mouth and grab its lip, but I'm not 100% sure. And I don't really want to take that risk if I'm wrong. So let's paddle it to the shore and we'll lift it out of the water."

"Kyle, it's huge! Are we going to make it? I only have an 8lb line on!"

"I don't know what else to do!"

Good thing I was in the stern so I could steer us towards shore. Randy delicately tugged the fish along with us. We found a mild slope, and he jumped out with his fishing rod. The fish was resting in the shallows, but as soon as we tried lifting it out of the water the line snapped and it swam away.

"NOOOO!!!!" Randy yelled.

"I am so sorry, Randy. I just don't know what to do with a catfish!"

"Let's try again."

So we did. Only this time, I had the bite, and my rod bent like nothing I'd ever experienced before.

After getting the fish close to the surface, Randy perked up.

"I have an idea! I'll use the leather oven mitt from the kitchen supplies!" Randy dug through the cooler and found the glove right as I was bringing up the catfish beside him. He reached down, fearlessly grabbed the fish's bottom lip like one grabs a human ear, and raised the fish in triumph. It wasn't as big as his, but we had landed our first catfish. From a canoe. It was glorious. We each took a picture holding it, smiling from ear to ear, before releasing it back in the water.

We were sad to leave, especially since we had just discovered that we could catch boatloads of catfish. But our families were expecting us home in a few hours. We loaded up the car, tied the canoe onto the roof, and started the drive south. But our joy didn't wane much, as shortly after starting for home we saw four moose cross the road, followed by a beaver the size of small bear.

Everything about our trip to the Kasakeemeemisekak was thrilling, and we promised ourselves that we would come back the following year.

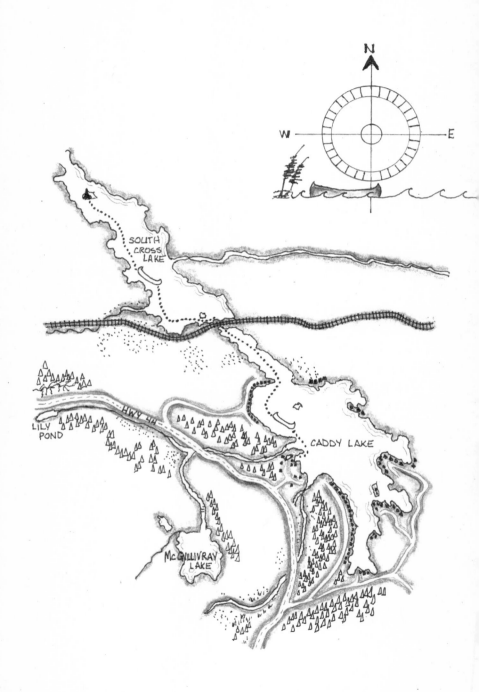

– 10 –

The Advantage of Working on Sundays

August 2016

Working most Sunday mornings means that getting away for a typical weekend of canoeing is quite difficult. I usually have to book the weekend off months in advance and end up being extremely selective about where I go paddling, and with whom.

However, one of the benefits of working on Sunday mornings is that I get Mondays off. So, as long as I'm willing to leave in the early afternoon on Sundays, I can squeeze in a quick two-day paddle.

This gives me a major advantage. Many of the popular canoe routes and campsites are quite busy on a weekend. And when Sunday afternoon comes around everybody packs up and heads home. Except for me.

Ashley was on maternity leave with our youngest child Milo, who was just four weeks old. Having a summer baby doesn't lend itself well to family canoe trips, and by the end of August I was pretty antsy to get back on the water.

I didn't want to leave Ashley at home with all three kids, as that would jeopardize not only future canoe trips but also my marriage. And possibly my life. So I offered a compromise.

I'd take Arianna and Zach with me. They were five and three years old at the time, had spent some time in a canoe with me, and loved camping. I'd plan a short route, remain close to civilization, and stay only one night. Ashley thought it was a great idea! When we pitched it to the kids, they jumped around and screamed for thirty seconds. They scurried off to their rooms and started packing their bags, even though we were only leaving in a week.

A short, close paddle for one night, and not wanting to portage everything by myself, meant that I'd return to my regular haunt of the Caddy Lake tunnels. The goal was to camp on the biggest island in South Cross Lake. That campsite is almost always occupied, but I hoped that the Sunday to Monday timing would work in our favour.

I checked the weather forecast and saw no rain during the day, but a lightning storm was predicted for the middle of the night. I figured we'd be in our tent by the time the storm hit, most likely asleep, so that didn't worry me much. A far greater concern to me was the wind.

There is an informal rule I have when paddling on South Cross Lake: Assume you'll be paddling into the wind. Always. You can find sheltered spots to avoid the wind on Caddy Lake, but South Cross Lake? It's just an open wind tunnel.

The forecast did call for wind, but much to my surprise the wind was supposed to be at our backs both days! I shared my excitement with Ashley, but, as she's paddled South Cross Lake a few times as well, we both agreed that it was a "believe it when you see it" situation.

I turned my attention to packing. I read in a canoeing magazine that one of the best ways to make your kids happy while camping is to give them each a small, waterproof bag and let them pack whatever they want, excluding food. So that's exactly what I did.

I gave Arianna and Zach each a dry bag. It only took a few minutes for the bags to be filled with colouring books, stuffed animals, and figurine toys.

I went into the garage and packed the rest of the gear that we'd actually need: tent, sleeping bags, sleeping mats, pillows, paddles, life jackets, water bottles, pot, buck knife, lighter, hatchet, plates, bowls, cutlery, bug spray, sunscreen, hats, sunglasses, ropes, tarp, folding chairs, clothes, towels, headlamp, and a watch. The funny thing about canoeing is that, whether you go for one night or seven, you end up packing the same amount of gear. The only difference is the amount of food you bring along.

At least the food wouldn't be too big a burden. Being the only adult, I made the meal plan as simple as possible. Hot dogs for supper, cookies for dessert, instant porridge for breakfast, and granola bars for snacks.

After church on Sunday, I threw all the food into one cooler and put the rest of the gear into the car. I called the kids to come and help me tie down the canoe. They mostly got in the way, but you have to start somewhere, right? We gave hugs and kisses to Ashley and Milo, and we were off for our adventure: Arianna, Zachary, and their dad.

When we arrived at the parking lot, my confidence in that island campsite being available skyrocketed. My assumption that everybody was coming off the water on Sunday afternoon was correct. The parking lot was bustling with activity as paddlers, fishers, and day-trippers were all removing their boats from the water. We did the opposite and loaded up the canoe. Zach sat in the middle seat and Arianna sat in the front. With smiles on their faces we pushed off!

They paddled enthusiastically for about three minutes. And then they stopped.

Well, more accurately, they stopped paddling forward and started paddling backwards. They thought they were hilarious and giggled between the two of them. When I stopped that nonsense, they started dragging their paddles in the water instead. I ended those shenanigans as well and told them that if they weren't going to help me paddle, the least they could do was not hinder me. We all agreed to this new arrangement, and they stopped paddling altogether.

This compromise worked because the wind was actually cooperating with us! It was quite a stiff wind, whipping up whitecaps on Caddy Lake. I wouldn't have made it far paddling solo. But with the wind behind me? Oh, we were flying. I needed to paddle just enough to keep us straight.

We passed other canoeists heading south, back to their cars. They were all on their knees, breathing hard as they dug deep. I smiled at one set of canoeists as we passed each other. I made sure to point out that I was paddling in the correct direction, and they were not. They laughed. I'm sure they were ready to heckle back, but they stopped paddling for a few moments when they saw the passengers in my canoe.

"Whoa! You're travelling with two little kids! Impressive!"

I smiled. "It's only working because the wind is in my favour. If the wind is like this tomorrow, I'm going to be in big trouble."

"Good luck!" they called, their attention now back to paddling into the wind and waves.

At the entrance to the first tunnel, I briefed my kids. No sudden movements. Keep your hands in the boat. And if we hit the side, Daddy will straighten out the boat. Their job was to just sit there. They promised to follow these instructions.

I blew my whistle to warn oncoming boats, and we let the current pull us along. The kids, like most people who travel through these tunnels, were mesmerized.

At the other end of the tunnel, we were greeted again by a tailwind. I could not believe my luck. All the close campsites on South Cross Lake were vacant, so I assumed that the island would be as well. This was working out perfectly.

I saw the island in the distance and started paddling. A few minutes later, Arianna turned around and asked, "Daddy, can I lie down?"

"Sure!"

"Me too?" Zach asked.

"Yup."

They both lay down, and within minutes the waves had rocked them to sleep. The dry bags filled with toys and colouring supplies were never even opened. They even slept through a loon swimming near us. And me? I was content to sit in the back and barely paddle, letting the wind and the waves do most of the work.

Even paddling lightly, we made it to the island in 75 minutes. I'd have given myself two hours to get there with a competent paddler in the bow. Perfect.

I roused the kids when we got close. They saw how close we were to the island and perked right up. We pulled up to a sandy spot and I jumped out, dragging the canoe as high as it would go. The kids hopped out and started exploring immediately. Trees and rocks and flowers and paths and fire pits and bugs and reeds and sand—Arianna and Zach were like kids in a candy shop.

It was at this point that I had to make a choice. I could either let them run around and explore the island on their own terms, or I could start teaching them how to set up a tent, make a fire, and launch them on their journey of being competent campers.

Both were good choices, but I chose option one.

Sure, it was more work for me, but when I stopped and watched them throw rocks in the water, or heard them laughing down the trail, or saw their imaginations running wild, I knew I had made the right choice. In fact, if them having this much fun in nature resulted in a bit of extra work for me, I considered it more of a gift than actual hard work.

After eating our hotdogs and cookies, we had a blast throwing more rocks into the lake. One of the best things about camping with kids is that, when they go to bed, you go to bed. We were in our sleeping bags, ready to sleep by 9:00 p.m. They thought it was a super late bedtime, as they rarely get to stay up that late at home, and I thought it was a super early bedtime, as I rarely go to bed that early at home. It was a win-win-win situation for all three of us. I let the kids tell stories as long as they wanted to, and ended up falling asleep in the middle of Zach telling a story about a puffin sleeping in a tent with a lion. I have no idea how that story ended, but, presumably, it didn't end well for the puffin.

I woke up to a flash of lightning. And another. And another. I looked at my watch. 3:00 a.m. The expected lightning storm was now here, and it was intense. I was amazed not only at how constant the lightning was, with at least a flash a second, but also at how quiet it was. This magnificent light show wasn't rumbling like I thought it would.

I leaned over and looked at Arianna on my left, and then Zach on my right. Their faces were lit up by the lightning, clear as day, but their eyes were shut, oblivious to the blinding light. I was thankful. For the next half an hour, when I wasn't checking to see if the kids were still sleeping, I was thinking about how I'd paddle against that wind if it didn't shift. No great answers came to mind. I finally fell back asleep, telling myself there was nothing I could do now. I'd just have to figure it out in the morning.

The kids were up early, as expected when camping. They picked up right where they left off, telling stories, swinging sticks, and jumping off stumps. I made some porridge and then started packing up. This time they helped me pack, and we were paddling by mid-morning.

My fears from the previous night were for naught, and my belief in meteorologists was restored. The wind had not only slowed down, but, more importantly, it had shifted to a northwest wind. We'd have the wind at our backs for the second day in a row!

The kids lazily dipped their paddles into the water until we got to the tunnel entrance. Now we were working against the current, so I gave them a different set of instructions. No sudden movements again, and Dad will still straighten us out if we hit the sides. But this time, you kids have to paddle and paddle hard. Do you understand?

Yes, they said, straightening their backs and tensing their faces.

Ready? Because here we go! Paddle!

They tightened their lips and furrowed their brows. I don't think those tunnels had ever seen a five- and three-year-old paddle as intensely as my two kids did that day. I J-stroked from the back, shouting encouragement, my face beaming.

I wrongly estimated the current, though. But in a good way! It was far weaker than I was anticipating, and with my two strong bow paddlers we easily made it through to the other side. My kids raised their paddles above their heads and cheered. So did I. Too bad we only had granola bars for a snack—they deserved something far better as a reward.

The memories of going on a canoe trip with their dad would have to suffice, however. And, considering they were asking for another trip before we got back onto the Trans-Canada Highway, I'd say those memories were more than enough reward for all of us.

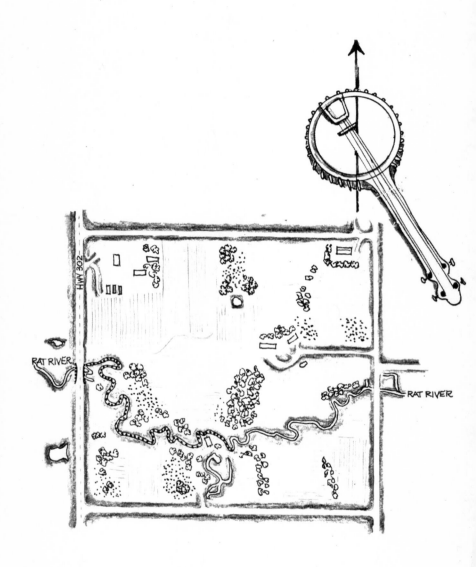

– 11 –

Paddle Faster, I Hear Banjoes!

April 2017

Winter is long in Manitoba. And dark. And cold. Getting through December is easy, with the joy and excitement and lights of Christmas. January and February are a tough slog. Especially with kids. Sure, you can bundle them up and keep sending them outside. But even then, when the day time temperature high is -25 degrees Celsius, being outside longer than twenty minutes is risking frostbite.

By the time March comes around, especially with Daylight Savings Time, I find myself staring out our living room window at the trees, longing for green grass, late sunsets, and open water for paddling.

We had an abnormally warm spring in 2017. The snow started melting early in March, and the sun kept sending down rays of warmth, comfort, and hope right through until April.

I started watching for the ice to break up on the rivers to expose navigable water. By the first week of April, the rivers were all clear. Arianna and Zachary, both a year older and wiser since our Caddy

Lake adventures, were determined to join me on my first voyage of the year. I extended the invitation to my friend Tom and his five-year-old son, Paxton.

Ashley got together with Tom's wife, Loretta, and they looked after all the kids who weren't coming canoeing—our baby Milo, and their triplets, Zander, Myelle, and Tasha (more on them in chapter 15)—while Tom and I pulled out my Backcountry Manitoba map book to look for the nearest river, preferably with as little civilization as possible.

There it was. The Rat River. Just north of the small town of Vita, it flowed west under Highway 302, the only roads near it being the mile roads. And, it was less than half an hour away!

We loaded up the canoe, put the three kids in the car, and drove down the quiet highway to the Rat River.

"Tom, when was the last time you've been down this road?"

"I have no idea. You?"

"Never."

In early April, nothing has started growing in most of Manitoba, so Mother Nature's colour palette is bland. Throw in a day with low, overcast clouds and the scenery is mostly brown, leafless trees, a mundane grey sky, and uninviting dark water flowing in the ditch.

We pulled the van over just past the bridge that spans the Rat River. To the right of the road was a little flat landing, perfect for loading and unloading. To the left was a farmer's field, accented only by the trees growing beside the river that ran through it. Near us were power lines with fish hooks hanging from the wires, left there years ago by aimless anglers.

We started paddling. The first point of interest was floating under the bridge. Of course the kids reached out to touch the support beams made of railway ties. And, of course, their hands became covered in oily pitch. And, of course, the water was about five degrees Celsius, so cleaning them up was no small task.

The cold water scared Tom and me. I've never capsized a canoe in my life, but having three kids in the middle, plus the high current from the spring melt, made me nervous.

Lucky for us, the spring melt was so fresh that the farmer's field was virtually one big lake. So instead of following the regular course of the river where the current would be the strongest, we were able to paddle almost anywhere on the field. We picked a clump of trees in the distance and aimed for them, stopping only to marvel that we were paddling over the leftover corn stalks from the previous harvest.

We came up to the trees and steered the canoe back into the main stream. Expecting the trees to be the only reprieve from the flat field around us, we were surprised to see an RV camper come into view.

"What the heck is that, Tom? Why is there an RV in the middle of a farmer's field?"

The river curved around the camper on three sides. It was eerily quiet. Its white colour was a sharp contrast to the greys and browns our eyes had grown accustomed to.

"Hello? Anybody home?" I called.

Nothing.

Tom called next. "Hello?"

Nothing.

"This is freaking me out, Tom. It's probably empty, but what is this? Whose is this? I can just imagine someone looking at us through one of those windows, drugs on the table, dead body in the closet, gun loaded, ready to shoot at anyone who steps near their camper."

The kids stared at it, sensing our apprehension.

"Who lives there, Dad?"

We kept paddling around the camper, noting the lawn mower, a stack of firewood, and some broken lawn chairs. Empty beer bottles were scattered everywhere.

"Probably nobody, Arianna. It looks like someone has driven a camper here and has made this their summer getaway. Like a cabin."

"Let's keep going, Kyle, lest we start hearing banjo music." We put our paddles back in the water with renewed vigour. Images of Burt Reynolds canoeing through hillbilly country in the movie *Deliverance* spurred us on.

The river narrowed, now staying within its banks a bit more. With the increased current we had to paddle harder. We encouraged the kids to dig deep and pull. Several trees were leaning over the river. We either deftly avoided them, or, if there were no branches, paddled underneath them.

After an hour of paddling we found a quiet eddy and took a snack break. Even though the sun started poking through the clouds, we were tired of paddling upstream. It was time to turn around.

On the way back, the current pulled us at tremendous speeds. When we encountered those trees leaning over the river, our margin of error was far smaller.

One tree, at least twelve inches in diameter, was leaning at a 45-degree angle over a curve in the river. Tom and I back-paddled as we lined up the canoe. I instructed the four of them to duck, and we'd let the current take us as I steered from the back.

"Ready? Duck!" Tom yelled. All four of them did.

What happened next seemed like an eternity to me.

I managed to steer the front half of the canoe under the tree, and everybody did what they were supposed to do. Perfect. It was now my turn to duck. By then the current had drawn the stern of the canoe towards the trunk of the leaning tree, which was now directly in between me and my passengers.

The solid tree trunk hit me square in the chest of my life jacket.

To avoid getting knocked into the frigid water I had three options.

I could try holding on to the tree and use myself as an anchor to stop the canoe and readjust. Given the speed of the current, I didn't like my chances.

I could jump onto the tree and hang on for dear life while the canoe drifted on without me. But water surrounded the trunk. Like a cartoon, I'd be stuck there until Tom came and rescued me. But Tom was in the bow, and there was no way that he'd be able to turn the canoe around, let alone paddle upstream without another competent paddler.

So I chose the third option. I lay down backward and leaned my head back like I was doing the limbo, praying that I wouldn't be knocked into the water. The tree bark scraped my chin as I floated underneath it.

I looked behind me at the tree. It had worked. The little passengers were excited about their ducking skills and were quite oblivious to what was happening behind them.

Tom turned around. "You alright, Kyle?"

I breathed deeply and shook my head in disbelief. "Yeah. But that was by far one of the scariest moments I've ever had in a canoe. If I had fallen into this frigid water, I'm not sure I would have made it back to the car before hypothermia set in."

An hour paddle upriver ended up being a fifteen minute return trip back to the car. And, even with heightened vigilance, we still saw no discernable movement when we passed the camper again.

Back in the van, we made a quick stop at the gas station in Vita. Some of the locals were having coffee and stared at us like we didn't belong there. We bought our ice cream and went home, happy for the early spring adventure.

That night, after tucking the kids into bed, Ashley and I heard whimpering coming from the kids' room. It was Arianna. Ashley went in and checked. I heard whispering, and then moments later Ashley came out and walked straight to me.

"Arianna told me that she's having bad dreams. Something about banjo music and people shooting your canoe and dead bodies. What happened out there?!"

"Oh boy. Yeah, that would be my fault. I'll take it from here."

I grabbed a pillow and lay on their bedroom floor until both kids fell asleep, aptly chastised for verbalizing my wild imagination. Next time, I'll keep my lips shut and just hum a banjo riff instead.

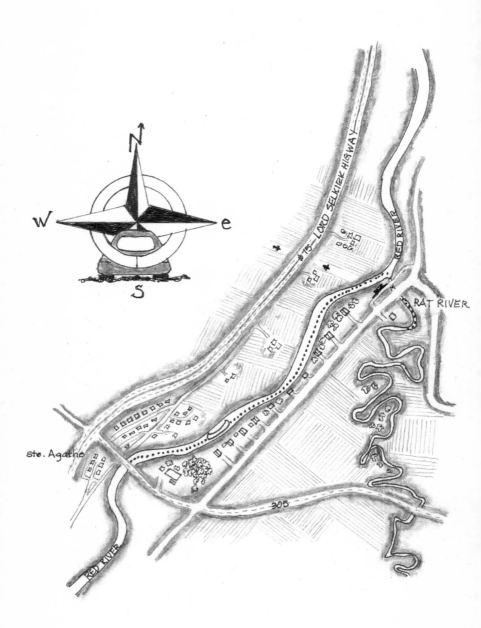

– 12 –

Taking Good Notes

June 2017

If you can get paid to paddle, you're drinking from the Holy Grail. You've reached the top, you've won your Stanley Cup, and you hang on to that job for as long as you can dip a paddle into the water.

I get paid to paddle.

Do I lead out-trips to remote rivers accessible only by plane? Nope.

Do I lead summer camps where I take kids out for extended camping trips? Nope.

Do I teach paddling seminars to new canoeists looking to upgrade their paddling skills? Nope.

I am a pastor. I work in a church. I work in an office. I sit in an office chair. And I work on Sundays, so a weekend paddle needs to be booked off months in advance.

How, you may ask, do I get paid to paddle?

We call it our "day-away strategy session."

I work with another pastor, Mel, who loves to canoe as much as I do. Every June, we book a day for long-term planning. We look back at the past year—at what we did, what worked, what didn't work, and look ahead to the next twelve months.

We don't always go canoeing for our retreats. Admittedly, it's hard to take notes while paddling. But who really cares about notes if you're paddling? One year we went to a restaurant, so we could take notes. We'll never do that again. Our notes were great, but there was no paddling. Other years we go to someone's off-grid cabin a few miles out of town. This is an acceptable compromise. We're in the forest and we light a fire in the woodstove. We go for walks down the trails, our thoughts wandering like the paths. There's even a small pond to look at. We can take good notes at the cabin, but there is no paddling in that pond.

No. The best day-away meetings are when we're in a canoe.

Several years ago we decided once again to get out the paddles for our day-away meeting. However, my daily schedule now included dropping off and picking up my kids from school, so Mel and I pulled out a map and started looking for nearby navigable water.

"Well, Mel, I've heard there are a few rapids on the Roseau River, and I don't think those will be conducive to taking notes. There's the Rat River near Vita, but that's just a glorified ditch with some sketchy drug dealers and dead bodies in campers."

"What? Dead bodies?"

"Yeah, well, what else do you think is in a camper surrounded on three sides by water in the middle of a farmer's field? Drugs and dead bodies. We're not going there."

Mel just looked at me blankly. I'm lucky to work with such a patient man.

"There!" I exclaimed, slamming my finger down on the map. "Let's get in touch with our Mennonite history and paddle to where the Rat River meets the Red River! That's where the first Mennonites landed in 1874 when they came here from Russia!"

Neither of us had actually been to this sacred shrine. What kind of Mennonite pastors were we? We wondered whether the historic site was north or south of the Rat River. We guessed south. We were wrong. But the road led us down a small hill, right to the river's edge, so we didn't make a big deal of not visiting the actual historic Mennonite landing spot. Maybe next time.

We took the canoe off Mel's Kia Rondo and put on our life jackets. The ground was dry and cracked near the river, so we avoided getting our feet muddy. And, like you always do when paddling a river, we started paddling upstream.

We were about 75 metres from a major highway, but it was as if we were in another world. Eagles perched in the large elm trees, eyeing a catfish or goldeye dinner from the Rat River below. Flocks of chickadees mobbed the eagles to keep them away from their nests. The gentle flow of the water, untouched by the wind, reached its last journey as part of the Rat River before joining up with the Red River, eventually making its way to Hudson's Bay. We saw deer prints embedded in the dry ground, which must have happened weeks ago after the last rainfall. Now I know how Frodo from *Lord of the Rings* felt when he had his blindfold removed in Lothlorien.

Several hundred metres up the Rat River, though, we were thwarted by a logjam. A huge, impassable pile of logs and branches thrust down there from the spring runoff. Determined to keep paddling, we got out to check how far we'd have to portage our canoe to bypass the logjam. The answer? About 50 metres. Quite doable.

Then we heard the dogs barking. We looked through the grove of oak trees and saw a house. The dogs were barreling our way.

"Mel, there is no way that I am going to put a canoe on my head while dogs are chasing us. Let's get out of here."

Thankfully Mel lived up to the nickname "Wiseman Mel" that I've bestowed upon him, and he readily agreed.

Back into the water and downstream we went. Satisfied that we were no longer a threat, the dogs turned around and went home.

We'd only been paddling for an hour, and still had many more notes to take. We could load up the canoe and find a different place to meet, probably a restaurant, or we could venture into the mighty Red River.

I'm delighted we chose the river.

The Red River here is about 50 metres wide. Being the middle of June, it wasn't the high point of the spring runoff so we could make decent headway against the current. Always paddle upstream first!

We paddled alongside massive river banks and were amazed at the high water marks colouring the trees, left over from previous spring floods. We saw farms close by, and the natural ditches formed when excess rainwater leaves the fields and enters the river. A few Canada geese swam in front of us, taking off in flight after we chased them for a bit. Despite the current, we were able to converse at a decent clip, reminiscing about the past and wondering about the future.

"How far do you want to go, Kyle?"

"I don't know. A few more kilometres? How far is it to Ste. Agathe?" Ste. Agathe was the nearest town, located right on the river.

I pulled out my phone and saw that we'd be in Ste. Agathe in about an hour, so that was the new goal. Having this objective certainly helped keep us focused when the rain started. After a few drops we threw on our rain gear. By the time we rounded the last corner and could see the town, it was pouring.

We paddled hard until we ended up under the Highway 305 bridge and out of the rain. We ate our lunches, listening to the raindrops pelting the river around us. Cars drove high above us on the bridge, oblivious to our existence. I have driven over that bridge countless times, but never once did I think there might be a canoe underneath it. And now I can't drive across it without looking for a boat in the river, despite never actually seeing one.

After copious amounts of note-taking under the bridge, the rain let up. We let the downstream current carry us northwards, putting our paddles in the water only to keep somewhat straight. It was peaceful, and we wondered why we hadn't launched our canoes in the Red River before today.

We paddled back into the Rat River and pulled up to shore beside the car. The previously dry and cracked ground was now slick and slimy since the rainfall, and our shoes were immediately caked in mud. After we tied the canoe to the car roof, we put all our wet and muddy clothes on a floor mat and started the car to head home.

However, Mother Nature had other ideas.

Mel hit the gas, and our tires started spinning.

"Uh-oh, Kyle. I'm not moving."

"No worries, Mel. I'll get out and push!"

I put my wet and muddy clothes back on and leaned into the hatchback. The car was on a grassy, flat area, so we made some headway. With me pushing, the car started moving. Mel pointed the car up the hill and accelerated.

Five metres up the hill, the car stopped. My feet slipped out from underneath me. I rolled out of the way as the car slid down the hill.

Huh, I thought to myself. *This is probably how people get driven over by cars.*

"Let's try again, Mel!"

Mel pressed the gas. I pushed with all my might. The car moved a few metres, and then rolled back down.

Mel got out of the car, looking up the small hill. "Shoot. What are we going to do?"

"No worries, Mel! We got this! We just need more traction. I'll start getting some branches."

Mel looked at me with kindness. "I don't think that's going to work, Kyle. We have to drive 50 metres up that hill. There's no way we can get enough branches."

"Well, we should at least start trying!"

I discovered that ripping branches off trees with your bare hands is much, much harder than using a saw or hatchet. I found a few small trees and a long, barky vine. I jammed them underneath the two front tires. Mel was skeptical.

"Come on Mel! A few metres at a time are better than no metres at a time!"

Mel humoured me and got back in the car.

"One. Two. Three. GO!" I hollered from the back.

The tires spun. I pushed. The vine sank down into the mud. The branches broke. The car didn't move.

Mel opened his door, looked at me and shook his head. "I told you that wasn't going to work," he laughed. "We're stuck."

"Well," I replied, "At least we have a warm car, a few hours before I need to pick up my kids from school, and cell reception. And last year I had to swim after my canoe in Lake Winnipeg, so we can be thankful that we don't have to swim down the Red River."

We sat in the car for a few minutes, going through our options. We could call a tow truck, we could call our wives to come pick us up after work, we could leave the car and try hitchhiking - none of these jumped out as great ideas.

"There's got to be a way out of here," I muttered, still determined.

We got out and looked up the hill at the road. It had two tire tracks, which were now filled with little rivers of rainfall flowing back to us. In the middle of the tracks was a raised mound of mud. On the left side of the road were shrubs and trees. On the right side, a few feet of grass, and then more trees.

Mel looked up the hill. And then behind us. And then back up the hill.

"We had traction on the grass down there, where it's flat," Mel reasoned. "Maybe we can get enough traction on the grass over there on the right?"

"It's worth a shot," I said. "What have we got to lose?" *Besides my life when the car rolls over me,* I thought to myself.

Mel backed the car up to the grass, aided by me pushing on the hood. Once there, I went around back and yelled, "Let's go!"

Mel hit the gas, and tires grabbed traction on the grass. Moving forward, he turned the car up the hill, aiming his left tires for the middle mound and his right tires for the grass.

It was working! The car moved faster than I could run in the mud, so I pushed off and stood off to the side, cheering. About ten metres past the point of our first attempt, the car slid off the mound and all four tires ended up back in the tracks. The car slid back to the bottom of the hill.

We cheered. This was doable.

We backed the car as far as we could on the grass, making sure to stop a few metres from the waterline. We figured that the car sliding back into the river would be the absolute worst outcome, right up there with me getting run over.

I heaved from the back. "Go! Go! Go!" I yelled.

The car gained speed. It turned up the hill. Mel lined up the tires and kept his foot on the gas pedal. The right tires grabbed the grass on the side, and the left tires stayed on the mound. I ran up the hill, following the car with my arms raised in the air.

"Don't stop, Mel! Keep going! You got this! GOOOO!!!!!"

At the top of the hill, where the dirt road became gravel, Mel stopped the car. He got out and looked at me as I dashed up the hill.

Shouts of joy went up as we embraced each other. We even spun in a circle, looking euphorically into each other's eyes with smiles from ear to ear.

"We did it! We get to go home without calling our wives!" we shouted.

What a great planning session the day turned out to be.

As I said earlier: If you can get paid to paddle, you're drinking from the Holy Grail. You've reached the top, you've won your Stanley Cup, and you hang on to that job for as long as you can dip a paddle into the water.

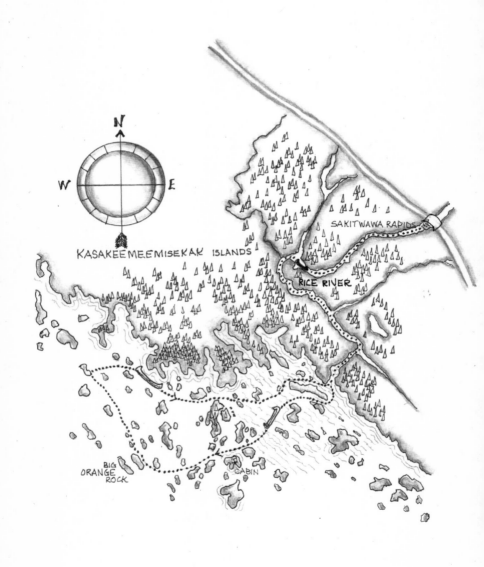

– 13 –

Aliens are Coming to Abduct Us

June 2017

Randy and I returned to the Kasakeemeemisekak Islands, but this time we brought along some old friends. Kevin and Phil had heard about our previous year's adventure and wanted to join us. I knew Kevin could paddle from the time we conquered the Mantario Wilderness Zone together. And I was pretty sure Phil had finally gotten over his fear of a Bigfoot leaving a ruffed grouse on the trail for him. Randy and I extended the invitation to them, and the size of our canoeing company doubled. We also figured that if our canoe drifted off again we'd at least have a second canoe available to go and retrieve it.

The weather forecast for our weekend trip was dismal. Despite it being the end of June, the daytime high was supposed to be twelve degrees Celsius, with rain expected on and off all weekend. But the worst part of the forecast was the wind. 50-60 km/h north winds were supposed to come down the lake, and the south basin of Lake Winnipeg had a water surge warning, with water levels expected to rise up to two metres.

Before we left, we had to decide whether to make this trip work or paddle elsewhere. Because we'd be fine paddling in the Rice River, where the wind and surge would be minimal, we decided to spend the first night on the river. From there we'd cautiously check out the islands, relying on the south-facing river mouth to protect us from the wind. And if we could find an island close to the shore to make camp, we would. If not, we'd duck back into the river and spend the day catching catfish.

I left Steinbach mid-afternoon and arrived at the Rice River around supper time, an hour before the others would arrive from Winnipeg. It was raining slightly, so I threw on my rain gear, left my bags in the car, and carefully carried my canoe and fishing gear down to the water. The river was several feet higher than the year before, so I paddled upstream as far as I could. I got further than we had the previous year. I steered myself into a large eddy and threw my line out.

Snag.

I paddled around my line, working to get my hook unstuck. After a few jerks, it was free. I tried again.

Snag.

This time my line snapped as I tried to free it. I tied on a new leader, jig head, and worm, and tried again.

Feeling a little tug, I pulled back to set the hook. Up out of the water flew a little smallmouth bass! While catching any fish is exciting, a baby smallmouth bass was not the fifteen-pound catfish I was hoping for. I let the bass go and rode the river's current back down to where Randy and I caught the catfish the previous year.

I dropped my line in and let it sink to the bottom. This time I felt a much larger, much stronger tug on my line. I pulled up to set the hook and felt an equally strong force pulling my line down. Catfish!

I played with it for a few minutes, reeling it in a bit, and then letting it pull my line out. In hindsight, I should have replaced my 8lb test fishing line with 20lb test fishing line made for hauling in catfish. But I took my time enjoying the experience. Despite the mistake of not bringing some stronger fishing line, I wasn't totally unprepared, as this year I had brought a net to get the fish into the boat.

Fishing solo for catfish in a canoe on a river is tricky, so the net was a life saver. I kept the catfish just below the surface with my fishing rod in one hand, and then scooped the fish from underneath with the net in my other. As the catfish thrashed in the net, I put my rod down in the boat beside me, and used both hands to lift the fish into the boat.

Success!

After pulling in three catfish in half an hour, the rain died down. I heard a car door slam, followed by enthusiastic laughter. I parked my canoe and met my friends up by the cars. They were ready to go, so we hauled down the rest of our gear and set off.

Within minutes three of us had catfish on our lines. The only reason we didn't have a fourth catfish hooked was that somebody had to paddle and help get the fish out of the nets.

We caught a dozen fish in less than an hour. Kevin and I, feeling like we should be a tad responsible, paddled a hundred metres down the river to our campsite and started setting up for the night. We had set up two tents, gathered some firewood, and started to prepare our evening snack when Randy and Phil finally joined us.

Kevin and I rolled our eyes. "Oh, look who came after all the work was done!"

They smiled at us, citing "catfish" as the reason they were late.

We stayed up a few more hours until the rain chased us into our tents. The nighttime low of six degrees Celsius meant we all went to bed wearing long johns and wool socks.

Upon seeing our breath in the morning we chose to keep those layers on. Phil reminded us all that there is no such thing as bad weather, only bad clothing. While technically correct, I still didn't warm up until lunch—despite paddling while wearing mittens and a toque.

We ate lunch at the mouth of the Rice River, interrupted every so often by the "whooshing" of an eagle flying over us. The sun was out and the water was calm—we had nothing to complain about.

We entered Lake Winnipeg with trepidation, ready to turn back if needed. We were buffered from the north wind by the shoreline. We found that, if we were willing to forgo efficiency and paddle in zigzags, we could hide behind islands. The waves were big, but not too big.

Phil had come to the Kasakeemeemisekak Islands last year with his brother Paul and some friends, Ryan and Reid. He shared a great story about how Ryan had grabbed his tent from the garage, and, upon opening the tent bag to set it up on an island, discovered that there weren't any poles in the bag! If you're new to canoeing and camping, be sure to make a mental note about this story.

Kevin had never been to the Kasakeemeemisekak Islands, so Randy and I steered the canoes toward Harvey Bushie's island. This year, we paddled to the dock side of the island. I also didn't bark like a dog while chasing pelicans, which was a wise choice, as again there were two fishermen from Hollow Water First Nation on the dock filleting their day's catch.

We paddled alongside the dock to greet them. After a bit of of small-talk, one of them looked at us rather seriously.

"Do you guys know that there's a wind and water surge warning today?"

"Yup!"

"And, you know that we're leaving here within an hour to go home because we don't want to get caught in that wind, right?" We looked at their 20-foot aluminum commercial fishing vessel with a 200-horsepower motor behind it.

"Good to know! We'll be sure to stick close to shore, and put our tent up high on an island!"

They shook their heads in disbelief and wished us well. We thanked them for their time and waved goodbye.

Despite our outward bravado, that conversation scared us silly. If the guys who drive on this lake every day were getting off the water, what the heck were we doing out here in our canoes?

The waters in and around the islands were still fine for paddling, but we shifted our number one priority from exploring the islands to finding a place to camp for the night. Given that our canoe had drifted off the year before when the water imperceptibly rose a few centimetres, we wanted to be extra cautious about a two metre surge warning. We started exploring the islands that were the highest above the waterline. We saw a few good camping spots on other islands, but they were too close to the water. We pressed on looking for higher land.

An hour of paddling and a few islands later, we found it. It was a bit of a steep slope to get to the top, but, once there, we were nearly four metres above the water. High enough to avoid any surge! We set up our tents, amazed at how they were bending in the wind.

Because it was only the middle of the afternoon, we jumped back into our boats with our fishing gear to see if carp were breeding again in that sheltered bay. Yes, they were breeding. And no, we still couldn't catch them. The cold weather meant they were lounging less along the surface, and they dove to the bottom as soon as we got close to them.

Kevin and I gave up first and turned back to our island. The wind was picking up, but we made it back safely. After hauling our canoe to the top of the island and tying it to a tree, we sat down on the rocks and waited for Randy and Phil.

We couldn't see them.

The wind picked up some more. Our tent flapped constantly.

We still couldn't see them.

Finally, in the distance, we saw them paddling hard. The wind was at their back, but if they pointed their canoe parallel with the wind they'd risk being whisked right past our island. But to paddle perpendicular to the wind meant they were getting hit broadside by the whitecaps. Kevin and I waited nervously, watching them being blown around like leaves in the wind.

"We should never have split up. We could have tied the canoes together into a flotilla."

Three pelicans were flying above their canoe. Well, actually, they were just hovering above their canoe, as the birds couldn't make any progress into the wind. After a few moments, the birds turned 180 degrees, and with the wind at their back soared out of sight.

Randy and Phil were making progress. They were both paddling on the starboard side to reach our island. The wind pushed their canoe broadside every time their double stroke straightened them out.

I leaned over to Kevin. In as quiet a voice as I could use while still being heard over the noise of our flapping tent, I commented, "Wow, Kevin. This lake does not care if we live or die."

Kevin looked at me, his silence an appropriate response. Randy and Phil kept paddling hard and finally came up to the island at the southern tip. They used some adept paddling skills to not capsize while trying to get out of their boat in those waves.

"That was just a bit of wind. No big deal!" huffed Randy, his feigned confidence masking the reality.

We looked as far south as we could see, to the islands neighbouring Black Island, seven kilometres away. With the open water, the waves were uninhibited for over fifty kilometres, and they were crashing against cliffs. White foam sprayed high into the air.

"Anybody want to paddle to Deer Island?" I asked.

"Yeah, Kyle. I do." Phil's voice dripped with sarcasm.

We set about making a fire pit for supper in a sheltered spot. After gorging on steak, potatoes, and fried vegetables, we sat in silence and watched the fire. The sitting turned into lying down. And the lying down turned into sleeping. After an hour, Randy woke us all up, not wanting us to sleep until dark. But with the wind blowing so hard even watching the sunset was exhausting—all we could do was hunker down by the fire and wait for the stars to come out.

And come out they did. Despite the sky being half covered in clouds, the stars we could see were glorious.

The wind finally started to die down. Around 11:30 p.m. there was an addition to the stars. A plane was flying towards us from the north, its wings blinking red and green. It was a large plane, and quite high, but we could still see it clearly. Where was it flying to at this late hour? I guessed that it was a medivac plane heading for Winnipeg from a northern community. But when it was above us, it turned around and went back north, out of sight. Strange, I thought.

Ten minutes later, with all four of us facing the fire, Randy looked up and pointed to the north. His eyes were wide open.

"WHAT IS THAT?!?!?"

A bright light was in the sky. The light was like a ball, quite stationary. Do planes have headlights? Was it a helicopter flying towards us? But we didn't hear anything! The light was so bright it was lighting up both the clouds and islands, creating a false sense of daylight. None of us had ever seen anything like this before. Maybe it was a UFO! We joked that it was aliens coming to abduct us.

I pulled out my phone to take a picture, but all I could capture was a blurry light in the distance. Determined to capture some sort of documentation, I switched to video mode and started recording.

"Okay, well, there is a bright light on the horizon, but the clouds are all lit up, and the light is reflecting off the water. We have no idea what it is, so if this is our last video, we love you, everyone. I love you Ashley, Milo, Arianna, and Zach." I hit stop. "If we die here tonight, I hope somebody finds my phone and can figure out my passcode."

Randy jumped in. "There's another helicopter! Or plane! Or something! There, on the left."

Phil squinted in the direction he was pointing. "That's definitely a plane."

The light slowly lowered on the horizon, and eventually went out. Darkness replaced the light. Before our eyes could adjust another ball of light appeared in the sky, identical to the first.

"That plane circled back, and this second light definitely came from that plane," Kevin said.

We watched the light slowly go down toward the horizon. This time, before it went out, the plane circled back again and another light appeared. There were now two shining balls of fire in the sky.

"It must be a search and rescue operation. And based on how slow those lights are falling, they must have parachutes." Phil said.

"I doubt it's a drill, as they'd probably pick a day other than the end of June to practice their nighttime drills," I said.

Kevin shivered. "I hope everyone is alright."

"Guys. That's why that plane flew over us a while ago. I bet it's the same plane! They probably saw our fire and thought we might have been the ones needing rescuing!" Randy said.

"Oh crap. Can you imagine if somebody out there needs rescuing, but instead of rescuing them, they came and rescued us?" I said.

We all fell quiet, thinking through how horrible that would have been.

We watched the lights for over an hour, each light taking several minutes to stop falling, always replaced by another one. By

1:00 a.m., it was over. We guessed that the lights were at least five kilometres north of us, so there was nothing we could do to help. We doused our fire and climbed into bed, saying a few prayers for whoever was stuck out there.

With only a two-hour paddle back to our cars the next morning, we slept in. We were making breakfast around 10:00 a.m. when we heard a motorboat approaching. We walked down to the shoreline and saw a commercial fishing boat coming from the south. It was being driven by two men: one of them from the dock yesterday, and a new one we hadn't seen before.

We waved. They waved back.

The man we didn't recognize called out to us. "You guys alright?"

"Yup! We're great! You?"

"Yeah! These guys came back yesterday saying there were canoeists out here! When we saw the flares last night we thought it might have been you guys, so we came out to check on you."

Our hearts overflowed with gratitude that these fishermen would boat twenty kilometres just to come and check on us.

"We saw the flares too! Do you know what happened out there?"

"No. Probably somebody's boat broke down or something. You guys going home today?"

"Yup. We're leaving right after breakfast!"

"Have a great trip!"

"You too!" They sped off towards Deer Island to check their fishing nets.

We found out afterwards that, somewhere north of us, a sailboat had capsized in the wind. The occupants had made it to land, but they were cold and wet. With the wind and frigid temperatures that weekend, we couldn't imagine how hard it would have been to wait to get rescued.

We slowly packed up our campsite and started to paddle home, grateful for the sun, warmth, and comparatively calm day. We pulled up a few more catfish at the launch point, and bid farewell to the Kasakeemeemisekak Islands until next year.

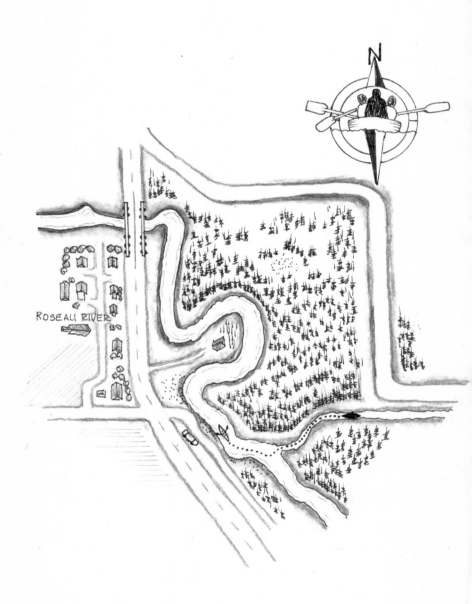

– 14 –

Me and My Boys

May 2018

With Ashley at work and Arianna at school, it was just me and my boys home on a Monday. We had one item on our list of things to do that day: Go grocery shopping. Of course we went canoeing first.

Zach, five, and Milo, almost two, "helped" me tie the canoe on top on our minivan, and off we went. Our destination that day? The Roseau River.

It was just a half-hour drive away. The goal was to paddle as far upstream as we could and simply drift back downstream to the car. It was the perfect plan for the three of us.

Two-thirds of the way there, I realized that I hadn't told Ashley where we were canoeing that day. In fact, I hadn't told anybody! I pulled over and saw that my cell phone had one bar of service. Phew. I sent her a text message, kicking myself for not being more responsible. That evening I asked her where she would have told the RCMP we were if we didn't come home. "I would have looked at them blankly. You could have been anywhere within an hour's drive."

I won't make that mistake again.

We pulled into the park overlooking the river and started unloading the gear. To get to the river from the parking lot you have to go down a steep hill and across a grassy area. To make less work for me I threw all the gear in the canoe and let gravity take it down the hill, with me just slowly walking beside it. The boys followed down the stairs, and we dragged the canoe towards the river until they got distracted chasing a garter snake that crossed our path.

I heard a motor nearby, getting louder, and looked up to see a three-wheeler ATV heading our way. I laughed. Those things were banned years ago in Canada for being too dangerous, so clearly this one was really old. As it got closer I saw that the driver resembled his machine—also really old. He was driving quite slowly, a tackle box and fishing rod hanging off the back. We waved and smiled at each other. He parked his trike under a tree while I continued to drag my canoe towards the beach.

I called my boys in from their snake hunting. The canoe was half in the water when the man came over.

"Hello!" he said.

"Hello! I see your fishing rod there. What are you hoping to catch?"

"Dinner."

"I see," I smiled. "What kind of fish are you hoping to catch?"

"Whatever bites my hook."

We laughed.

He looked at my canoe, my kids, and then back to me. "Where are you off to today?"

"Oh, just taking the kids for a paddle. We're going to paddle upstream as far as we can go, and then drift back."

"Oh, good luck with that," he said slowly. "The river's pretty high, so you won't make it very far. Plus, there are some rapids just around the corner that'll give you trouble. Your best bet would be to paddle downstream and have somebody come pick you up."

"Shoot. That sounds like a great idea, but it's only us three, and we only have one car. I guess we're stuck paddling upstream from here."

"Oh well, good luck then!" he said. "I'll be here fishing if you need anything."

"Good luck to you too! I hope you catch your dinner!"

He went back to his tree to put a hook on his line. I finished loading up the canoe and gave instructions to my boys.

"Milo, your job is to just sit here in the middle and not move, okay?" He nodded.

"Zach, your job is to sit in the front and paddle with me, okay?" He nodded.

We set off. The current quickly caught my bow and turned it downstream, but I countered with a few strong C-strokes and got us pointed back upstream.

"Okay, Zach! Paddle!"

He tried. He really did. But five-year-olds aren't effective paddling partners when trying to work upstream on a river with high water levels. We crawled forward, passing the fisherman after a few minutes. He waved. I smiled and nodded, not willing to give up any of my hard-earned river real estate by taking a hand off my paddle. I found a slower moving corner of the river along the far riverbank and we finally made decent progress. We made it two hundred metres upstream, turned a corner, and immediately all our forward progress was halted. The river was straight from there, with no curves for us to hide behind. The current hit us with full force.

I called for Zach to paddle harder, and, despite his determination, we still didn't move. We weren't going anywhere anytime soon. I turned the boat around and decided to come back a different day with a stronger paddler in the bow.

Just before turning the corner and seeing the fisherman again, I saw a small tributary flowing from the east into the Roseau River. As we had been paddling for only ten minutes and had hours to spare, I ducked the canoe into there to see what we could find.

We found a beach with hundreds of minnows that scattered each time our paddles hit the water.

A bit further into the tributary we found turtles, sun-bathing on fallen trees. We snuck up on one and I captured it. My passengers were impressed.

Still further in, we found a family of ducks. The babies couldn't fly yet, so all they could do to avoid us was swim further upstream. This was quite comical to my kids. The ducklings eventually hid in some tall grass off to the side, probably thankful they were no longer being chased by a 17-foot canoe.

There were little sections of shallow, babbling water where I had to get out and pull the canoe over the rocks. We paddled under a few trees, taking in the small sand cliffs on each side of the tributary. The sun shone down on us, and we were happy to go at whatever pace we wanted, stopping to look at anything that drew our attention.

Eventually I realized that we were paddling in a glorified drainage ditch, as the trees were being replaced by fields and the curvature of the river was straightening out.

We turned the boat around and drifted back downstream, ducking under the fallen trees and scaring our family of ducks again.

Right where the tributary flowed into the main river I beached the canoe. We made sand castles. We threw rocks into the river. We worked hard trying to catch one of the minnows. Zach was disappointed that we hadn't brought one of our dipping nets. I agreed, as these fish seemed to be smarter than your average fish; I couldn't chase or trick them into swimming into one of the sand traps I made.

"Hey Zach, do you know why these fish are so smart?"

"Why, Dad?"

"Because they go to school!"

My dad jokes were not much better than my minnow-catching skills that day.

After an hour of playing on the beach, our stockpile of snacks had been consumed. Plus, we technically still had to go grocery shopping that afternoon. We jumped into the canoe and paddled across the river back to the original beach.

When we passed the old man fishing, he had fallen asleep under his tree. One of Zach's paddle strokes clunked the side of our canoe and startled the fisherman. He waved at us.

"How far did you get?" he asked.

"You were right! Not very far! But we had a good time on that beach over there. How's the fishing?"

"Oh, still waiting." He was still waiting for his dinner when we left the park.

Zach and Milo fell asleep on the ride home. My thoughts drifted to how I'd need a stronger second paddling partner to get upstream.

But, halfway home, a piece of wisdom floated by—I reminded myself to be fully present in the moment. I should be aware of what was in front of me, instead of already planning the next trip.

I was thankful for our small adventure. I was thankful for the time spent in nature. I was thankful for the blue skies and gentle wind. I was thankful for the minnows and the turtles and the ducks. I was thankful for the time in the canoe. And I was thankful for the day spent with my boys.

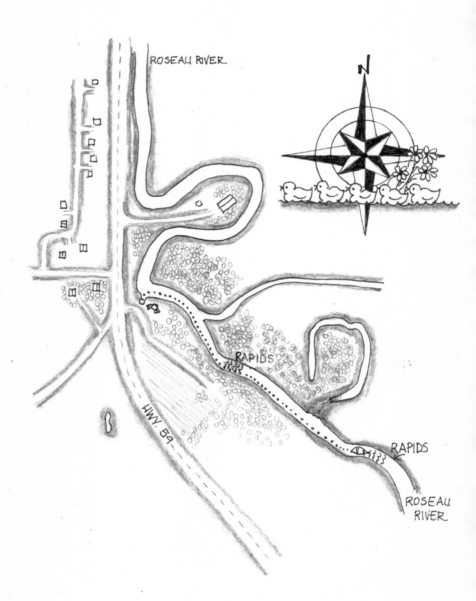

– 15 –

Seven People in a Canoe

June, 2018

When you win the lottery, you obviously make changes to your life. You might quit your job and take up a hobby. Like building a canoe, for example. You might move to a fancier house, one with a built-in irrigation system or an indoor pool. You might go on a dream vacation to that luxury hotel in the South Pacific where the floors are made of plexiglass and you can see the fish from your bedroom. You might go back to school, pay off your debts, or buy a new car! The specifics don't really matter all that much. What matters is that you won the lottery, and your life is never the same.

This is exactly what happened to our friends, Tom and Loretta. They won the lottery.

Only they didn't win any money. They won a different kind of lottery: the lottery of life. Instead of having a second child after Paxton, they had their second, third, and fourth child, all on the same day. Yes, they had triplets!

Zander came first, followed by identical girls, Myelle and Tasha. The odds of this happening? About 1 in 10,000. All of their friends were amazed that the odds of having triplets are far, far better than the odds of winning the actual lottery.

Regardless, they won the lottery of life. And like every other jackpot winner, they made major life changes. But instead of quitting her job to build canoes, Loretta quit her job as a teacher to change eleven poopy diapers a day. Instead of buying a bigger house with an irrigation system and an indoor pool, they bought a bigger house so they could fit all the cribs into one room. Instead of a bucket-list vacation to that island far away from their friends and family, their friends and family moved in for the next two years of their lives. And instead of buying a posh new car, they bought a used Dodge Caravan in order to fit all those car seats. And the colour of the van? "Old man beige." In hindsight, the Dodge Caravan purchase was the antithesis of what we expect when someone wins the lottery. But "winning" triplets is definitely just as life-changing.

When they were given the news that they were having triplets, the ultrasound technician wisely made sure that Tom and Loretta were sitting down. A few days later, Loretta invited Ashley over to tell her in person. When she got there, she slid into bed with Loretta and held her tight. "Let's do this." And since then, we've done our best to raise our kids together.

Okay. Back to paddling.

After I had attempted paddling the Roseau River with my boys and realized that I needed another adult to actually paddle upstream, I asked Loretta and her just turned four-year-old triplets to join me and my boys next time. Because what can go wrong with five kids under the age of six in one canoe?

I sent Arianna on the bus to school, loaded the canoe onto my van roof, and packed the paddles, lifejackets, rain gear, and snacks. I buckled in Zach and Milo and off we went to pick up our friends.

WHO GOES CANOEING WITH THEIR MOTHER-IN-LAW?

We pulled into their driveway, and the triplets were giddy with excitement. Zander had already buckled himself up in his own car seat in the Dodge Caravan. He had to be convinced that, yes, we were still going canoeing, even if he had to get out of his van so we could put his car seat into my van. Myelle and Tasha insisted on packing their own bags. They had each picked a bright, sparkly pink purse to hold their version of canoeing essentials. Myelle's bag was filled with trucks, while Tasha stuffed hers with all sorts of pink and purple trinkets.

As Loretta and I were in the kitchen doing a rundown on all the food we had packed, I pulled out my phone to look at the weather forecast.

Rain.

Darn. I don't mind paddling in the rain, but with five kids?

I checked the radar forecast from my weather app.

"Look at this Loretta. It looks like it's going to rain now, and then in one hour there will be about a 90 minute break. That's our window. Are you still game?"

I could see Loretta quickly doing the mental math. How would the kids react if the canoe trip was cancelled? How did she feel about being stuck inside all day with the kids? But would canoeing be worth it if everyone was wet and miserable the whole time?

"Let's do this. And if we only get a few minutes on the water, we'll call it a successful outing."

What a great friend I have.

We drove thirty minutes through the rain to the landing spot, watching for breaks in the clouds. On the way, we went over some basic canoeing rules with the kids:

1. Be Brave.
2. Have fun.
3. No sudden movements.
4. Stay in the canoe.

When we got there, we parked under a tree and waited a few minutes.

The rain lessened.

"Let's start getting ready."

Putting rain gear on five kids isn't a quick and efficient task. Especially when life jackets have to fit over all the layers. The kids ended up looking like five blobs with no necks and T-Rex arms sticking out.

The rain lessened.

We took the canoe off the roof and made the kids wait under the trunk door of the van.

The rain stopped.

"Let's go."

We threw our snack bags into the canoe and, with five kids following us like little ducklings, dragged the canoe to the water's edge.

We put Zander on the canoe floor in front of me behind the yoke. We squished Zach, Milo, and Myelle on a removable seat in the middle. They were crammed in there, but then at least they couldn't fall out. Tasha was stashed in the bow by Loretta's legs.

A quick note about Tasha. Being brave wasn't exactly her strong suit. She was scared of most things, including the friendly neighbourhood dog. We knew she might not do too well paddling up and down the small rapids of the Roseau River, but we also hoped we could push her a bit to learn how to triumph over her fears. So, while one of the kids had to sit in the bow entirely because of space restrictions, Tasha ending up there was quite intentional on our part.

Loretta and I hopped in and pushed off from shore.

My calculations were correct. Having a strong paddler in the bow was enough to make headway against the current, even with our cute cargo secured all over the canoe. Especially since their rain gear made them quite the useless paddlers. Milo also refused to relinquish the one kid-sized paddle that he was clutching awkwardly, so nobody even had a chance to try without getting in his way.

We paddled around the corner, beside some sand cliffs, and under a tree hanging over the water. We startled some ducks, and watched them fly away. It was remarkably serene as the birds started singing after the rainfall. And, hey, when you're the parent of small children, you're usually happy with the ten seconds of serenity you get just from closing the car door and remaining outside the vehicle after you've buckled in your kids. So when you're in a canoe with them and everybody is quietly watching the birds fly by, you enjoy every moment of it.

It was short lived, though.

We rounded a corner and came across a set of Class I rapids. Small waves stretched across the river, standing still with white tips. We picked our route and paddled.

After a few strokes, we were spit right back out. The river here was only a few feet deep, so the speed of the current had picked up dramatically.

We picked a different route and tried again.

Same result.

The third time, though, I was determined to get up these rapids. Right before the river pushed us back downstream again, I jumped out into the shallows. Perhaps I should have told everyone in the boat what I was doing first, as some of the kids were quite concerned about me. So I waded up to the bow of the boat and reassured them.

"Hang on, kids. We're getting up these rapids. If we can't paddle up them, then we'll walk up them!"

Zach and Zander started laughing at me in the water with my jeans rolled up to my knees. Milo and Myelle quietly watched. Brave Tasha whimpered in the front, while Loretta whispered words of confidence into her ears.

A few minutes later we had passed the rapids, and the rhythm of paddling settled upon us once more. We noticed patches of blue sky poking through the clouds in the distance, and hoped that our little window of rain-free tranquility would last.

It did not.

At least the rain was gentle. We rounded a bend in the river only to be greeted by another set of Class I rapids. Because these weren't as shallow, getting out and dragging the canoe wasn't an option.

We picked a route and jammed our paddles into the water.

The river sent us back.

We tried again.

The river was having none of it.

Tasha's crying got a bit louder.

I steered the canoe to the edge of the river.

"Grab onto that clump of grass, Loretta. I'm going to get out and see what's ahead."

I left our little adventurers in the care of Loretta, clinging to some grass, and started bushwhacking to see what lay ahead of the rapids.

More rapids. Bigger ones. Small, in terms of how big rapids can be. But big enough that I didn't want to tackle them with seven people in a canoe, no matter how small some of those people were.

I looked up at the sky. Grey, with no more breaks. Our window was over. It was time to turn around.

On my way back to the canoe, I noticed little white flowers growing near the ground. I picked five of them for our five little canoeists. The children beholding a little bit of nature's beauty was the consolation prize for having to turn back sooner than anticipated.

I'm always amazed at how much faster one paddles downstream than upstream. And fast we went.

So fast, in fact, that I missed our preferred route through the shallow rapids. We hit a rock, got spun around, and started heading downstream backwards! I tried turning the boat around.

But by the time we got broadside, we were in the really shallow part and got hung up on a gravel ridge.

The four kids in the middle were clinging tightly to the thwarts and gunwales. Loretta was torn between helping me paddle and ensuring that Tasha's increasingly panicky cries and attempts to climb on her lap weren't going to lead to any erratic movement. We could hear the raindrops splashing around us, getting louder every minute. I grimaced at our predicament. With his free hand, Milo was still clutching his paddle.

"Alright, kids. I'm going to jump out again and push. Just hang on and we'll be back at the van in no time."

I jumped out. Without my weight in the back, the stern lifted. I grabbed the gunwales and let the current straighten the boat out. I walked as far as I could on the gravel ledge, instructed Loretta to stick her paddle in the water as an outrigger, and jumped back in.

We were free.

We got back to shore and loaded the canoe onto the van. We found a picnic table under a tree and enjoyed our snacks out of the drizzling rain. We played with the trucks from Myelle's purse, happy to be able to stretch our legs on solid ground. The kids didn't seem disappointed by the very short canoe ride, or too rattled by our white-water rafting.

On the way home, the kids fell asleep talking about how much fun they had, wondering when we were going to do this again.

After we had all gone home and tucked our kids into their own beds, I received a text from Loretta.

"These were Tasha's words after the trip: I love, love, LOVE canoeing. Except for the swirly water. I was a bit scared at that part. But I still LOVE canoeing!"

Don't we all, Tasha? Don't we all.

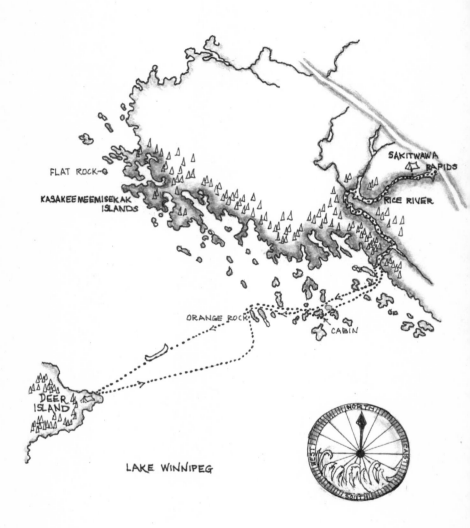

− 16 −

Windbound

June 2018

We added two more people to our now-annual canoe trip to the Kasakeemeemisekak Islands. Jeff and Nick were joining Randy, Phil, Kevin, and me this year, bringing our number up to six.

The five of them lived in Winnipeg and couldn't get off of work until 4:00 p.m. on Friday, so we made plans for them to meet me at the boat launch at Sakitwawa Rapids. I packed my small Honda Fit full of gear, tied the canoe to the roof and left my home in Steinbach at 2:30 p.m., figuring I'd get there for 5:30 p.m. and have two hours by myself to fish. I was right about having two hours by myself, but I was very wrong about what time I'd get there.

I turned off Highway 304 onto the Bloodvein Road expecting a smooth 30-kilometre gravel road to the Rice River. Only a few kilometres up the road, though, I was stopped by a man wearing a yellow hard hat and neon orange vest, holding a stop sign.

I rolled down my window. "Good afternoon!"

"Good afternoon!" he replied, swatting the horse flies away. "Where you from?"

"Steinbach. You?"

"Manigotagan. Just down the road from here."

"So what's going on up here? I thought the road was in great shape. It's fairly new, isn't it?"

"Yeah. Just making it wider, and building a better base. They're expecting more traffic in the future so we're starting to fix it now."

"Cool. How many kilometres are you doing now?"

"Eleven."

His radio beeped. "Hang on a second," he said to me, and lifted the radio to his mouth.

"I'm here, with someone waiting to get by. What's going on over there?" I couldn't understand the person on the other end of the radio, but he clearly could. He spoke into the radio, "Okay then. I'll let him go when you're done the dynamite blast. Let me know when. Over and out."

"Dynamite blast?"

"Yeah. Instead of trucking in rocks for the base, we're just making our own. We're using dynamite on the cliffs beside the road."

He stopped talking and scanned the length of my Honda Fit, all thirteen feet of it. His eyes moved up to the 17-foot canoe strapped on top.

"You think you can make it with this small car?" he asked, a smirk on his face.

"Well, I made it the last two years just fine. But you tell me. How bad is the road ahead?"

He paused. "I bought a big Toyota Tundra truck a few months ago, knowing I'd be working on this road. But, after a week, I realized that it would get wrecked up here. So I bought an old Dodge Caravan instead, so now I don't care when it gets trashed on this road.

I stared at him, mouth open. He looked at my car again and smiled. His radio beeped. He flipped his sign from "Stop" to "Slow."

"You can go now. Good luck!"

"Thanks!" I said, closing my window.

I have never, in my life, driven through a worse construction zone. They could have shot a commercial for ATVs on that road. Bumps and ruts and sudden drop-offs were everywhere. There were backhoes and dump trucks around every bend. When they saw me, every worker along the way stopped working and laughed, pointing at my Honda Fit with a canoe on the roof.

After a few slow kilometres, the surface of the road switched from gravel to six-inch rock base. I drove even slower. And then it switched to ten-inch rock base. I stopped the car and got out. With the horseflies buzzing around my head I looked at my tires and their fifteen-inch diameter, and back to the road on which I was supposed to drive. *You have got to be kidding me*, I thought.

I drove even slower. I pulled over to the side to allow two oncoming dump trucks to get past, their tires as big as my car. When they were clear I had to rev my engine to get enough speed to get my tires back onto the ten-inch rock base.

But I made it. Until the ruts stopped me, that is.

Up ahead a worker had stopped oncoming traffic, and two large pick-up trucks waited. Both the worker and the drivers were watching me with smiles and looks of disbelief on their faces. In between their vehicles and mine were ruts, at least 45 centimetres deep. I stopped my car and got out, and was again greeted by horseflies. I looked at the deep ruts. I looked at my car. If either side of my tires slipped, my car would bottom out and probably get hung up.

I looked at the high points in between the ruts. I looked back at my car. I went to the ruts and stretched out my arms in an effort to measure the distance between them. I went back to my car and stretched out my arms in an effort to measure the distance between my tires. I looked back at the road ahead of me, and the trucks waiting for me to pass by.

I could make it.

I lined up the left side of my car with the ruts on the left so I'd be riding on the high spots. And off I went.

Oh boy, was I nervous. I stuck my head out the window to see if my tires were in fact driving on the high points of the ruts. They were. I was going to make it!

When I passed the construction worker and the waiting trucks, I fist-pumped out my open window, a smile plastered to my face. They tipped their hats with respect and admiration for what that little car with a canoe on its roof had just done.

After driving on ten-inch rock base, the next section of six-inch rock base was easy. And when it turned into two-inch gravel, it was so smooth it felt like I was skating on ice.

I knew I had made it when I passed the "Construction Zone Ends" sign. The final ten kilometres to the Rice River flew by, and soon I was in my canoe catching catfish, waiting for my friends. An hour later than planned, but none the worse for wear.

We had texted each other before I left cell phone range, so I knew they were about ninety minutes behind me. And since one of their cars was also a Honda Fit, I knew they'd travel as slowly as I did. I imagined what they'd say when they saw what kind of shape the road was in.

"Oh my goodness. What is this? Did Kyle make it through here?"

"What if we come across him on the side of the road, waiting for us because his car broke down?"

"He'd text us if he turned around, so I guess we just keep going, assuming that he's there."

And that's what they did. I heard them talking from the parking lot, right when I expected them. I reeled in my fishing line and went up to greet them.

"Hey! You made it!" both parties exclaimed to each other. We shared stories of our drives. I was correct in expecting that they'd keep driving, assuming I had made it. By the time they got to the road, though, all the workers had gone home for the weekend. My friends couldn't even ask if a blue Honda Fit with a green canoe on its roof had driven by! They didn't recall any ruts, so I guess a backhoe had come and flattened out those crevasses for other cars. The construction workers had left all the heavy machinery on the side of the road, and my friends were eager to show me the pictures they took of themselves climbing into the buckets of the backhoes.

"Hey Kyle," Phil said. "What would you have done if we hadn't made it? Or if we had turned around?"

"Oh yeah. I was ready. I knew you were about an hour and a half behind me, so if you weren't here within two hours, I was just going to pack up my stuff and head home. I'd look for you on the side of the road, or wait for a text message when I got back into cell range, explaining where you were."

"Smart plan. Now let's go catch some catfish."

The six of us carried our gear down in no time, and within twenty minutes each boat had a catfish on someone's line. Nick and Jeff couldn't believe it was this easy. After an hour of catching catfish we paddled down to the campsite. Night was fast approaching, so we set up our tents and sat by the fire for hours sharing stories.

The next morning was calm, sunny, and beautiful. Some of us made breakfast while the others caught more catfish. Kevin ended up catching a pickerel, our first one in three years of trying. I filleted it and stashed it in a cooler for that day's supper. After eating, cleaning up breakfast, taking down our tents, and putting on sunscreen, we set off down the river.

We paddled the river at a slow pace, alternating between casting our fishing rods and seeing how close we could get to the eagles perched in the trees.

Along the way, we saw a cow moose and her baby. Nick and Randy, who were in the lead, turned the corner first and saw the two moose wading in the shallows. They beckoned to Phil and me to stop talking, and in turn we shushed Jeff and Kevin. The two moose looked up at us and quietly walked into the forest, leaving only their footprints behind.

"See! This place is so cool!" I said. "I love paddling to the Kasakeemeemisekak Islands, and we're still in the Rice River!"

Even though the route was familiar to a few of us by now, turning that corner and seeing the islands in Lake Winnipeg still brought out so much joy. Because the weather was cooperating we steered straight out into the lake, our only goal being to find a place to stay for the night.

But first, we needed lunch. We pulled our canoes up on an island with a rocky edge and started eating. The sun was beating down on us, so Randy and Phil kicked off their sandals and waded into the water.

"OWWWWWW!" Randy yelled. "Something sliced my foot!"

"Mine too!" shouted Phil.

"Guys! It's the zebra mussels!" I kept my sandals on and walked out to where they had found a spot free of zebra mussels.

"Zebra mussels?" someone called from the shore.

"Yeah. Let me see if I can find one." I reached down into the lake to find a rock with zebra mussels on it. "OW!" I yelped, pulling my hand up. Blood dripped down my thumb. "Those little buggers are sharp!"

Zebra mussels are an invasive species originally from Europe. They were introduced to the Great Lakes in 1986 via the bilge water from an ocean liner, and in 2013 they were introduced to Lake Winnipeg. They likely hitched a ride on a boat. They're the size of a fingernail, can reproduce by the millions, have no natural predators here, cling to every rock surface, and eat so much algae that they permanently change ecosystems. And they are as sharp as a knife.

The three of us carefully got back to shore to tend our wounds, amazed at how much they bled. Compression, elevation, and bandages barely slowed the bleeding. We explored around the shallow parts of the rocky island, and, sure enough, there were zebra mussels everywhere.

"Wow. These things are on every rock in this lake. Unreal." We knew that we'd have to swim with our shoes on from here on out, and made a mental note to look up how to ensure our canoes wouldn't transport the mussels to new bodies of water in the province.

We paddled by Harvey Bushie's island again, and, like the previous years, there were the fishermen, filleting their day's catch. Flocks of swimming seagulls and pelicans surrounded them, and we rather enjoyed watching the hundreds of birds take flight as we paddled near them. As we approached them to say hi, we saw a third person with them, sitting on a picnic bench.

"Hello!" we called.

"Hello!" they called back.

"Great weather today, huh?"

"Yup."

"Excuse us, but we have a question for you. We see you're filleting fish there. Is there any chance we can buy some off of you?"

"Yup! Come on up!"

I paddled my canoe over to some rocks and Phil hung on while I jumped out, both of us now on the lookout for zebra mussels. "Thanks," I said, shaking their hands. "We only caught one pickerel back there, so this is great. By the way, my name is Kyle."

The older man at the picnic bench stood up. "I'm Harvey Bushie."

"Harvey Bushie! So nice to meet you! You have a really nice place here!"

He thanked us and offered us some fish for five dollars a bag. Luckily I had come prepared for this exact situation and passed over a twenty dollar bill, amazed at how cheap this was compared to grocery store prices.

He asked us where we were going. We said that we didn't know, but if the weather stayed like this, and the wind stayed calm, we might paddle across the open water to Deer Island.

"Deer Island?" he said, looking at our boats. "Be careful. We only go out in the open water if there's a second boat nearby. You know, in case our boat breaks down or the wind picks up. You don't want to get caught out there in the open water."

"Yeah, thanks for the heads up. We'll be extra careful!" I was starting to notice a trend over the years.

I looked out towards Deer Island, and then back towards the fishermen. I reached out to shake their hands again. "Thanks for everything! Have a great day!"

I climbed back into my canoe and the six of us took off, looking forward to adding fresh fish to our steak dinner.

We paddled to the big orange rock and joined up in a flotilla to discuss what we should do. Our options were to paddle northwest, towards the campsites from our previous years, or to head straight west and hit the beach on Deer Island. We scanned the horizon, saw a few clouds gathering far in the south, and turned our canoes toward the open water.

An hour later we were running along the beach, marveling at how we had this place all to ourselves again. And the best part was that zebra mussels can't attach themselves to sand, so there was nothing stopping us from running around barefoot!

We fished, but, without any terns to scout the water for white bass, we came up empty. We swam, digging our toes into the soft sand. Near some fresh moose prints we found a tiny creek of cold water flowing into the lake, the water chilled by permafrost.

We kept an eye on the clouds in the south. They were getting bigger, darker, and had patches of heavy rain. But they were mostly heading straight west, nowhere near us. The clouds growing in the west were white and fluffy, and the wind was still calm, so we weren't worried. We did decide, though, that we'd make a fire, eat a late lunch, and then paddle back before any of the nice fluffy clouds turned angry.

We dug a pit and lit a fire. Kevin brought out the sausages from his cooler, while the rest of us found rocks to place around the edge of the pit so we could balance our grill over the flames.

As we built up the fire the wind shifted from the east to the south. It gained strength, picking up bits of sand and flinging them against our skin. The clouds covered the sun. We put on our shirts, and three of us ran to the canoes and paddled behind a sand bar, keeping our boats out of the waves.

While Kevin tried to cook our sausages, the rest of us gazed out at the southern sky. Walls of rain held up the dark clouds like giant pillars. And the pillars kept getting bigger and closer. Waves crashed over the beach where our canoes had recently sat.

We posed for pictures, yelled at the rain, and guessed if the storm would hit us or miss us. Finally, we agreed that the rain indeed was going to hit us, so we had to make a plan.

Looking past the beach, we peered through the trees. The forest was far too thick, plus the wind and the waves and ice had deposited a wall of sand at the tree line and covered it with deadfall. A tarp or a tent wouldn't work in there.

A quick glance around the rest of the beach told us there weren't many other options, so we worked with what we had. Fearing a surge in water levels, we set up two tents as high on the beach as we could. The canoes were dragged to the tree line, flipped over on top of our gear, and then tied to some trees. The wind was getting stronger, the rain approaching so fast that Kevin gave up on making lunch. He stashed the half-cooked sausages in his cooler and entered into the fray of prepping for the storm. We searched our bags for playing cards and our beverages of choice and jumped into the tents to wait out the passing storm.

We called between the tents, and agreed that we would quickly pack up and paddle back across the open water once this squall had passed. This was probably one of those brief, summer storms that dropped a lot of water in a short period of time but is usually followed by calm.

It's worth noting here that Randy, Nick, and Kevin had set up their tent properly and were safely settled into their tent, awaiting the oncoming storm. Phil, Jeff, and I thought it was just going to be a short rain squall, so we only set up the fly of the tent, stretching it out over the poles. This seemed like an efficient idea at the time as it gave us more time to tie down the canoes. But, when the storm came, we realized that nothing was stopping our tent from flying away like a kite! We each grabbed a pole with one hand, and Jeff grabbed the fourth pole with his foot. We were dry, but quite unable to play cards or enjoy our drinks.

The storm came, and it came hard. Rain pelted down on our tents like machine gun fire. The wind flapping the tents added to the noise. Puddles started forming around the edges of our tent, slowly seeping through the sand to where we lay. I looked out the fly window, and couldn't see the four kilometres across the lake to the islands. Everything was blocked out by a grey curtain.

But we were safe. And mostly dry. The falling rain on our fly acted as white noise, lulling us to sleep as we lay there.

Forty-five minutes later, we woke up to Randy, Nick, and Kevin standing outside our fly.

"Guys. Wake up!"

We crawled out from under the tent fly and surveyed our surroundings. All the canoes were still there. So was our gear. It was cloudy now, and colder, but as expected the rain had passed by and was now travelling north. The only thing that we were wrong about was the wind. It hadn't died down one bit, and was still gusting straight from the south.

Our hearts sank. We knew that we couldn't cross the open water with such a strong wind coming from the south. We'd never make it with those waves pounding us broadside. Even if we did try to make it back to the mainland, and miraculously not capsize, diagonally tacking in and out of the waves would take hours.

A new plan was formed. We'd have supper here on Deer Island. If the wind died after supper, we'd paddle back to the Kasakeemeemisekak Islands and camp on Harvey Bushie's island. The mid-summer sun meant that we'd have light until 10:00 p.m., so we had at least a few hours to wait for the wind to weaken.

Jeff fired up the cooking stove, and I set about battering, coating, and frying up the fish as an appetizer. All four bags of fish were devoured, with Nick claiming this was the best fish he had ever eaten. I told him it was easy to make pickerel taste so good when it had been filleted only a few hours earlier. The warm food in our bellies lifted our spirits.

The rain had put out our lunch fire, so we relit it and put on the steak and potatoes. The steaks had been marinating for days and were oozing with goodness. We feasted until we could eat no more.

We watched the setting sun, then turned our faces to the south. We were running out of daylight, and the wind was still howling. Grimacing, we made yet another new plan. We'd spend the night on Deer Island, and then leave at the break of dawn to get back across the open water. The wind usually dies down at night, so we set our alarms for 5:30 a.m. and hoped for the best.

We cleaned up our food, set up our tents properly for the night, and threw our bags inside. With a few hours to spare, we collected firewood for the largest bonfire Deer Island had ever seen. When we were done making a four-metre-high teepee of firewood, we sat down on a sand dune and waited for darkness.

As the stars came out, we still could not believe that the wind speed hadn't decreased one iota. It was still so windy it took the six of us twenty minutes to light that pile of wood on fire, as our lighters and kindling kept being blown out. Using our bodies as shields from the wind, it finally lit. But the wind pushed the fire over, and it started to burn horizontally. Sparks carried far into the distance. At least we were smart enough to build the fire far away from our tents so the sparks couldn't put a hole in them.

Discouraged about our turn of luck and unexpectedly being windbound on Deer Island, we clambered into our tents for night. Lying in my sleeping bag, I went over what our options were if the wind didn't die down.

We could risk paddling in it, which wasn't a great option. If one of the canoes capsized, a T-rescue would be almost impossible. And having three grown men in each of the other two canoes was risking further disaster.

Our second option was to paddle around to the north side of Deer Island, out of the wind, and then paddle west to Gull Harbour. This wasn't ideal either, because then we'd be on the other side of Lake Winnipeg with no way to drive the five hours around the south basin of the lake back to our cars. But at least we'd be off Deer Island and able to call our families.

The third and final option was to wait until the wind died down. But I knew of stories about people being windbound on Lake Winnipeg for days. My sister-in-law Kira had led the Camp Stephens six-week canoe trip, and, even after paddling for eight hours a day, running rapids, and doing one hundred portages, those girls still had to spend three days stuck on Lake Winnipeg, waiting for the wind to die.

Plus, for some reason, our phones weren't picking up the cell signal that we had a few years ago, so we had no way to communicate with our families. I could just imagine the panic that would set in for them if we didn't return when we were supposed to.

I didn't like any of our options and prayed that the wind would stop. I fell asleep to the noise of my tent flapping in the wind.

I woke up every hour that night. Before I could even check the time, I'd hear the wind. It wasn't dying. I prayed some more.

My alarm went off at 5:30 a.m. Randy was already standing outside my tent.

"Guys. It's still windy."

My heart sank even further.

I got out of the tent and looked around. The sun was rising over the horizon straight east of us, right where we wanted to go. Blue skies everywhere. And wind.

Kevin joined us. "But look! The wind shifted! It's no longer a south wind. It's coming at us almost straight from the east!"

The wind still pelted our faces, but the waves were smaller than the night before. I guess the wind only having a few kilometres to whip up high waves was better than the wind having many kilometres to whip up even higher waves.

We huddled together.

"So, what do we think?"

"I think we do it."

"It's now or never. We can't really expect that wind to die down during the day."

"We might be stuck here for days."

"Paddling straight into the wind is slower, but it's far better than battling broadside waves."

"And if we find that it's too rough, we can always turn around and ride the waves back to this beach."

"Okay then. It's settled. Let's skip breakfast and break camp as quickly as possible. We paddle close together. And we wear our life jackets."

Jeff piped in. "Great. Kevin, you brought me a life jacket, right?"

"Nope. Randy did, right?" Kevin asked.

"No. I thought Kyle brought an extra one for you."

"No, I thought you guys had it all figured out," I said.

Jeff's face dropped. "So, I don't have a lifejacket?"

Silence. Here we were, about to cross the open water of Lake Winnipeg on a windy day expecting waves to crash over our bows. None of us were overly confident, and we only had five life jackets for the six of us? Seriously?!

"Well, Jeff, you're the only one of us who's a lifeguard, so you're the best swimmer. It has to be you."

He obliged. What else was he going to say?

We packed as fast as we could, the psychological stress creating pits in our stomachs.

We pointed the boats straight into the sun and loaded them up.

"Alright, guys. Paddle close together. We got this. Straight east, into the sun."

We put on our sunglasses and pushed off.

The waves were breaking on the shore, so each paddler in the bow had a wet start. Nick was Randy's bow paddler, Jeff was Kevin's, and Phil was mine. We quickly got to deeper water where the waves were breaking less. We paddled on in silence.

I aimed the canoe for the sun and counted my strokes.

"Fifty strokes a side, Phil, and then we switch. You set the pace, and I'll follow your lead."

"Got it."

And so we paddled. Randy and Nick to our left, and Kevin and Jeff to our right. Every few minutes we'd check in with each other. All was well. Someone would shout when a wave crashed over the bow, with significantly more shouts coming from our right than our left. Kevin and Jeff were in Kevin's old aluminum canoe and soon started to lag behind the newer, lighter canoes.

As we paddled further into the open water the wind shifted slightly towards the north. We shifted our direction to match it, now paddling a bit south of where we wanted to go. With the waves trying to push my boat broadside, halfway there my J-stroke could barely keep us straight anymore.

"I can't keep us straight, Phil! I need to paddle on the right side only. I'm sorry to tire one side out, but I'll holler if I can switch."

"Whatever you need, Kyle, just let me know. We're getting there!"

And we were. When we had travelled a few minutes and hadn't had to turn back, we felt much better about our chances. It would be hard work, but, unless some disaster occurred, we expected to make it across. In the rhythm of paddling, and the seriousness of the moment, time flew by. I like to think that the image of our worried and borderline panicked wives texting each other that night about which of them was going to call for a search and rescue operation drove us to paddle harder than usual.

Finally, the sun rose high enough to no longer backlight all the islands. They changed from being large black shapes rising out of the water to islands with rocky shores and green trees. I saw the big orange rock to our left.

"There's the orange rock, you guys! That's where we want to go!"

So we turned our boats to the left and felt the waves carry us towards the big orange rock. We were going to make it!

We gathered out of the wind behind the big orange rock to assess how we were doing. My boat was relatively dry, with maybe only one wave crashing over Phil. A few more waves had hit Nick in the front of Randy's canoe, but they were good. The aluminum canoe, though, had at least an inch of water in the bottom of it. Most of the waves they encountered left their mark, both on Jeff and in their canoe.

"Whoa. That's a lot of water in your boat," I said. "Geez. Good thing we put the guy with no life jacket in there."

We agreed to head over to Harvey Bushie's island to have our breakfast, and dodged the wind by paddling behind islands. We didn't stick together as much here in the islands, as each canoe was trying to find the shortest route. When we were alone, Phil turned around and looked at me.

"Well, we made it."

"Yeah. That wasn't really much fun, was it, Phil?"

"No. No, it wasn't. If I'm honest, I don't think I would have even attempted to cross that water if I was paddling in the back. But I knew you've paddled a lot in your life, so I figured we'd make it across."

"Thanks, Phil. That means a lot. And right back at you. Because as nervous as I was this morning, I was glad to have a strong paddler in the front."

We docked our canoes at Harvey Bushie's island, calling to see if anybody was there. But of course it was currently uninhabited, as they would have seen the storms the day before and gone home. The rest of the guys threw together a skillet of eggs, bacon, peppers, mushrooms, and Kevin's half-cooked sausages from the day before. I lay down under a pine tree and slept, my nerves finally calming down. They kindly woke me up with a hot plate of food. I have good friends.

"I don't think I'll be paddling to Deer Island again any time soon," Randy said. The rest of the paddlers agreed.

We navigated the last little bit of windy Lake Winnipeg until, relieved, we got into the calm waters of the Rice River. We paddled slowly, enjoying the easy paddle after the morning we had. To further release the tension of the day we hopped out of our canoes and spent half an hour climbing the Sakitwawa Rapids. Our laughter at Nick when he fell into a small pool half way up the rapids and discovered a monster catfish swimming towards his legs further calmed our hearts. Except for Nick's, that is.

We tied our canoes to the roofs of our cars, loaded up our gear, and took off for home, right back through the six-inch and ten-inch rock base layer of the Bloodvein Road. This treacherous road felt easy compared to the harrowing journey we had completed a few hours earlier. But, on the way home, we affirmed that neither the road nor the wind would deter us from returning to the Kasakeemeemisekak Islands next year.

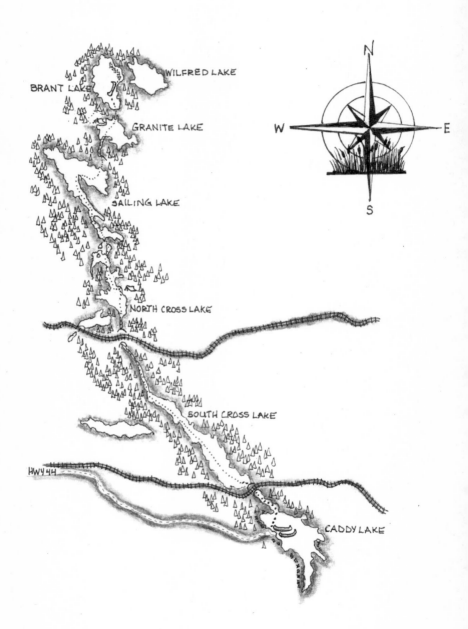

– 17 –

What Do You Mean You Don't Like Portaging?

July 2018

"Are Crocs okay?" I double checked the text message to make sure I was reading it correctly.

I was.

I had sent out a text to all the people coming on this canoe trip. "We'll be portaging. Bring shoes."

"Are Crocs okay?" was not the response I was expecting. *What's the worst that can happen?* I thought. At least they didn't ask if flip-flops were okay. So I texted back, "Sure!"

Our annual camp in Pauingassi First Nation had been cancelled. Their entire community had been evacuated to Winnipeg because of a forest fire. My leaders, Samantha and Alyssa, and I decided to offer something to the volunteers who suddenly had a free week in July: a canoe trip.

But not just any canoe trip. We dreamed up a canoe trip that most of them wouldn't plan or experience on their own: a canoe trip into the Mantario Wilderness Zone.

It had been ten years since my last foray into the Mantario Wilderness Zone. It was the perfect trip for our group! We were young enough to handle some decently long days, it was remote enough that we'd probably be alone, and we would do a couple of portages to make us feel like it was a tough trip. The aim was to come back sore and tired but proud of what we accomplished. And with some good stories to share, of course.

Sam and Aly are smart, tough girls with lots of camping experience but limited time paddling. We weren't dumb enough to think that taking a bunch of rookie canoeists and campers deep into the Mantario Wilderness Zone was a good idea. Rather, the plan was to paddle for four hours on the first day and camp on the island in North Cross Lake, then spend most of the second day paddling and portaging to a great campsite on Wilfred Lake. Day three would be about getting back to North Cross Lake, and then we'd have a leisurely paddle back to our cars on day four. Plus, I had made the trip from Caddy Lake to Wilfred Lake in one day a decade ago, so surely we could make it in two.

I was right, but barely.

There were ten of us in total: Sam, Aly, Patrick, Mike, Regan, Arthur, Naomi, Connor, Michael, and myself. We were aged 19-35, and ready for adventure. We picked up the rental canoes early in the afternoon and were at the boat launch by 3:30 p.m. We had a four-hour paddle ahead of us, but, since we had packed an ample amount of snacks and could count on the late sunsets in early July, we were okay with a late supper.

We got to the boat launch on Caddy Lake, loaded up the boats, covered ourselves in sunscreen, and paired off.

I checked if they all knew how to steer a canoe, and everybody said they knew enough to sort of maintain a straight line. If they didn't know how to J-stroke, they told me they could at least rudder, and that was a good enough start for me! I mentally planned to come alongside the ones who were having a hard time staying straight and teach them as the trip progressed.

The first three boats took off. I watched the stern paddlers. J-stroke, rudder, rudder. Perfect.

The fourth boat took off.

Circles.

Okay. That one will need a bit of work. Michael was in my bow, so we paddled up to the circling canoe.

"You got this," I said. "Your job as the front person, Naomi, is just to paddle. You're the engine. Pick a side, take twenty, thirty, fifty strokes, and then switch. And do it again. You can't steer from the front, so don't even try. Just put that paddle in the water and pull." She nodded, informing me that she had paddled on her school's dragon boat team that year, and those skills would transfer quite well.

"Regan, you're in the back, so you're the one who steers. If you want to steer the canoe right, paddle on the left side. If you want to steer the canoe left, paddle on the right side. But the key is not to switch sides every two strokes, as then you won't go very fast or very straight. Instead of switching sides, or lifting your paddle out of the water right away to start a new stroke, just keep it in the water for a few seconds. Your boat will straighten out, and then you can resume paddling. Got it?"

"Got it." They started paddling.

Circles.

I smiled. "Okay, Regan, you need to anticipate the boat turning and start dragging your paddle before your boat has turned. It's not a car, where the wheels are on the ground the whole time and they respond immediately to the steering wheel. Your canoe is floating on the water with momentum, so any movements have to be anticipated. Try dragging your paddle every other stroke to get a feel for it, okay?"

"Okay."

Circles.

I looked at the other three canoes far off in the distance, already halfway to the first tunnel.

"Let's try again. You got this!"

Circles.

We had spent thirty minutes paddling as far as I could usually paddle in five minutes. I wasn't opposed to teaching, but I knew we still had a long paddle ahead of us. I didn't want to cook supper in the dark, so I'd try one more time. I explained what they should do, I pulled up beside them to show them what I meant, and then backed off to watch them go.

Figure eights. Not much better.

I was out of ideas. Luckily, Naomi wasn't. She turned around abruptly. "You know what? This isn't working. Mike, do you know how to steer a canoe?

"Yup."

"Then let's get to that island over there and switch. Mike and Kyle can paddle in the back, and we'll paddle in the front. Deal?"

I looked at Regan. "Only if you want."

"Yeah," he said. "Otherwise we'll never get there. You can teach me later."

"Deal."

So Michael and I paddled straight to the island, and the other canoe took its sweet time circling its way there. But they arrived, eventually. We switched paddlers and everyone was happy. Off we went to catch up to the others.

We passed through the tunnels as usual. I've been through them countless times, and, while I still love paddling through them, this time I loved watching the faces and hearing the reactions of my friends paddling through them for the first time. Silence. Awe. Observation. Wonder. The slowing pace of the paddles. Ducking from the birds. Reaching up to try to touch the roof. Cameras snapping pictures. It never gets old.

We paddled through the second tunnel, grateful for water low enough that we didn't have to duck, and came out into North Cross Lake.

North Cross Lake is beautiful. It's small with rocky shores and lots of bays and inlets to explore. We had been paddling for a few hours now and our arms were getting tired. The setting sun offered stunning backlight to the other canoes, while also frying the left sides of our faces. We paddled around a fallen pine tree leaning over the water and spotted the island we'd be camping on.

After making landfall we divvied up the tasks. We sent some people to collect firewood, others to set up the tents, while Sam and Aly prepared supper for us. We sat on the edge of the island to watch the sunset, told stories around the fire, and went to bed tired but content.

I woke up to a brilliant flash of lightning, followed by a massive crack of thunder. The sun was rising, bringing a thunderstorm with it. A few more bolts of lightning left their marks across my retinas, but I rolled over, content to let the rain splattering our tent put me back to sleep.

When the rain stopped, we ambled out of our tents. Despite it already being 10:00 a.m., we were in no rush. Some of us made bacon and eggs for breakfast, while others played cards in the tent. It was wonderful.

In an effort to be efficient, Patrick and Mike took their backpacks out of their tent and put them by the canoes. Then they got sidetracked by breakfast. As I was walking past their now-empty tent, a wind gust lifted it up and sent it tumbling towards the water. Thankfully a tree stopped it from tumbling into the lake. When I saw Mike and Patrick eating their breakfast, I informed them about the precarious state of their tent. They sheepishly put down their plates to go rescue it.

We left the island at noon, and soon came up to the small portage around the dam into Sailing Lake. When we hit the large open water on Sailing Lake we turned east to find the portage into Granite Lake.

I had done this portage twice in my life, and was looking for the big yellow triangle made of two-by-fours with the orange tips and lettering on the bottom. But I couldn't see it. I had a rough idea of where the portage was, so I headed in that direction. Closer to shore, I was surprised to see that in the place of the wooden triangle there was now a small white metal sign, similar to a "No Parking" sign. It was bolted to a tree, and in red lettering it read *Granite Lake: 1000m.*

We ran our canoes aground. I got out and started packing up the loose items in my canoe.

"What do we do now?" someone asked.

"We portage," I said. "We're not in a rush, so let's do this in two trips. The two light canoes can be solo-carried, the other three can be double-carried. And the other two people can carry our coolers and paddles and tents. And then we can all come back for our bags. Let's go."

And we did.

On this particular portage, I discovered something profound. Despite being one of the oldest people on the trip, and possibly the most out of shape, I didn't mind portaging. In fact, I was surprised by how much I was enjoying it. Sure, it's hard work and the sweat drips into your eyes and you can't swat the bugs because you're carrying a canoe over your head. But there's something rewarding about setting your mind to a hard physical task and completing it.

I made the two trips, the first one with a canoe on my head, and the second one with my bag and some random gear that we hadn't properly attached to our bags. After dumping all the gear, I walked back to help out the strugglers and the stragglers. I met Connor and Arthur near the end of the portage, standing by their canoe taking one last break. Since they were close to finishing, I moved on until I met Sam at the halfway mark making her second trip with a cooler in tow.

"I'm here to help!" I said, taking the blue cooler from her arms. We started walking together.

"Honestly Kyle, I was just about to lose it. I know I'm pretty weak but that cooler almost seemed heavier than the canoe." I popped the lid open to take a peek. Staring back at me were five bottles and one four-litre jug, all filled with water. No wonder this thing was heavy. I was quite impressed that Sam's noodle arms had made it this far.

Sam and I ended up passing Arthur and Connor, who were still taking their break, and met up with the rest of our crew. We waded through seaweed and sunken logs until the water was up to our necks. After a few minutes of cooling ourselves off we were back in our canoes paddling across Granite Lake to the next portage.

It was a short paddle, and after passing a few beautiful campsites we were under the next white sign that read *Brant Lake: 500 metres*.

We emptied our boats and pulled them up on shore. I put a bag on my back, a cooler in my arms, and led the way.

The path was easy to follow at first, but it quickly disappeared into a tangle of junipers. I tried remembering my last Mantario Wilderness Zone trip, but over the past decade all the portages had blended into one. I searched for an orange arrow that would mark the way, but found nothing. I did find a few stacked rocks clearly created by humans, but they had been there for ages and were now covered in lichen and vines. We followed those for a minute, but they abruptly stopped after 100 metres.

I couldn't believe that there were no other trail markers. Those white signs weren't there the last time I had paddled here, so somebody had to have been here. But there was no visible evidence.

I put down my cooler and started trailblazing. The rest of my motley crew had caught up to me, and without words had decided to wait until I had found the path. I scampered over trees, through junipers, down rocks, and there it was! I saw a familiar orange arrow, nailed to a now-fallen tree. I pushed through some willows and found the trail, now overgrown with grass.

"Found it! Follow me!" I called as I picked up my cooler and started pushing through. The corridor through the willows led to a rocky opening from where we could see the lake. This part triggered my memory, and I knew instinctively to crank a hard right, walk through the tall grass, and straight to an abandoned beaver lodge.

Seeing me trudge through the tall grass and onto a beaver lodge, somebody shouted from behind, "Is this the right way?"

I looked up and saw the white sign. *Granite Lake: 500m.*

"Yup!" I called back. I put down my cooler and bag and headed back to Granite Lake for a canoe. As I passed Aly on the way, I remarked to her, "'I'm fairly certain we're the first ones here this year, but I don't know. Based on how overgrown everything is, I'd even suggest that maybe we're the first people who have been here for several years!"

And that's the key to backcountry canoeing. Most canoeists won't portage at all. Some canoeists will portage once. Only the die-hards, or, as Bill Mason said, the crazies, will portage more than once. But if that's the price you have to pay to find pristine wilderness, so be it.

Not everybody on our trip agreed with me. We had been paddling and portaging for eight hours already, and we weren't even at our campsite yet. We had donated our blood to mosquitoes, black flies, horse flies, and deer flies. I plucked ten wood ticks off my legs after one section of this portage alone. I was in my element, loving every minute of it. I had even created a few backcountry canoeing converts. But I was losing some of the others. Especially those who were trying to portage in Crocs.

I did my best to rally them. "We're almost there! We just have a short paddle down Brant Lake, and then a short carry to our campsite. We can even leave our canoes here on this lake. And trust me, it's a beautiful campsite! It'll be worth it!"

Everybody rallied enough to at least pretend that they were having a good time, and we were back in the water by 6:30 p.m. We paddled north with a strong wind at our backs, looking for the portage. The wall of wild rice that impeded progress on my last trip had yet to grow high, so we paddled right through it. We were so close to our day being over, our stomachs being filled, and our bodies resting in our sleeping bags.

Ding!

The electronic noise invaded my serenity. And it came from inside my own canoe.

"What was that?" I asked my paddling partner.

"I got a text from my girlfriend!" he exclaimed.

"What?!?" I said. "We're in the middle of a lake in the middle of the bush, and you're texting your girlfriend?"

"Hey!" he fired back. "I've worked hard! I've done what I'm supposed to do! I've given it my all! I'm tired and hungry and want this day to be over, so excuse me if I'm texting my girlfriend! I miss her!"

I looked around and assessed our circumstances. The portage should be close, the wind was pushing us in the right direction, and I had pushed some of the other canoeists harder than they had ever been pushed before. So, I didn't stop him from texting his girlfriend.

I think I chose wisely for two reasons. One, the reception was terrible, so only a few texts would have made it through. And, two, I had a much bigger problem on my hands.

We couldn't find the white portage sign.

We paddled up and down the shoreline. Everything was overgrown, and there were no indicators of a launch point. We pulled out the map and compass. We knew we were close, but saw nothing. A few of us even got out of canoes and started walking along the shoreline looking for a sign, an arrow, a pile of rocks, a remnant of a path, anything. Nothing.

Aly, Sam, and I huddled together.

"Well, we can't find this portage, and there are no campsites on this lake. We have two options. We know that Granite Lake is 250 metres straight east, so we can blaze trail until we find the lake. Because the campsite is big and rocky, we can't miss it. And from there, we can hopefully find the actual trail and walk back here."

We looked into the dense, darkening forest that we'd have to bushwhack through.

"What's the other option?" Aly asked.

"We paddle back to the beaver lodge, portage our gear back to Granite Lake, and spend the night there," I said. "It's 7:30 p.m. now, so we still have two hours of solid sunlight."

"Both options sound terrible," Sam said.

"Yup. I'm mad that we can't find this portage, but at this point, it is what it is," I said.

We went back and forth between the options. One option was so close, but if we didn't find the campsite, or the trail back, we'd be in deep trouble. The other option was so much more work, and turning back would be demoralizing. But we at least knew that at the end we'd have a place to cook our food and lay our heads down.

A bird in the hand is worth two in the bush, so we chose option two. We pointed our canoes south, into the wind, and started paddling. While the wild rice wasn't big enough to slow our pace, it was big enough to keep the waves down, so that was a small blessing. The only blessing. It wasn't our longest paddle, but the wind beating against our already-exhausted limbs made every stroke a challenge. I was pretty sure most of the paddlers were cursing my name at this point.

My canoe got to the beaver lodge first. We quickly unloaded, hauled our gear and canoe up to the rocky clearing, and went back to help the next four canoes do the same. Once we were all in the clearing, we made a plan.

Eight of us would start carrying backpacks, coolers, kitchen supplies, and tents back to the campsite. Sam and Aly would start cooking, Connor and Regan would get firewood, Michael would filter water, and Patrick, Mike, and Naomi would set up the tents. Arthur and I would flip all the canoes over, lean them against each other, store the paddles, life jackets, and my fishing rod under the canoes, and follow behind with the last two bags.

The plan worked flawlessly. By the time Arthur and I got to the campsite there was a fire, clean water, tents were going up, and supper was being prepped. We were starting to resemble a real backcountry canoeing outfit!

We ended up sitting down for supper at 10:00 p.m. that night, our frustration, anger, and disappointment slowly disappearing. We acknowledged the terrible parts of the day, but all found something positive to focus on. We laughed at the stories of wood ticks and beaver lodges. I worked hard to convince them that, yes, that portage did exist, and that I hadn't led them into the middle of nowhere just for fun. And we relished that the distance we had to paddle and portage the next day just became that much shorter.

Before we went to our separate tents that night, we all climbed into the six-person tent to play cards and tell stories. Mike admitted that he was genuinely afraid of bears because, when he was camping as a kid with his family, a black bear had come right up to his tent looking for food. The bear left him alone, but the memory never faded. To ease his anxious soul, we sang "Everybody (Backstreet's Back)" by Backstreet Boys as loud as we could. Any nearby bears would have run the opposite direction once they heard how awful we sounded.

Maybe, just maybe, I'd even have a few more backcountry converts by the end of this trip.

We woke up to a bright sun in a blue sky. Since we had a bit of a shorter day ahead of us, we slowly made breakfast, content to leisurely sip our coffee. Sitting for breakfast, we talked about the plan for the morning.

"We have three tents to be taken down here, and a campsite to pack up. And, we have five canoes still at Brant Lake, plus life jackets and paddles and my fishing rod. Carrying canoes will be the worst job, so who's coming with me?" I looked around.

Regan, Patrick, Connor, Arthur, and Mike all volunteered. We started walking, and helped each other locate the wood ticks stuck in our leg hair. Mike, Patrick, and I each soloed a canoe, while Regan and Connor took another. Mike and I agreed to walk back one more time to get the last, heaviest canoe. Arthur said he'd carry all the lifejackets, paddles, and my fishing rod.

And carry the boats we did. At this point in the trip portaging had become less of a dreaded horror and more of a resigned necessity. It involved hard work, lots of sweat, and some grunting and cursing, but when we lowered the canoes off our heads at the end of the portage we felt quite accomplished and strong.

But the real highlight of this portage was walking back to get the last canoe with Mike. I didn't know him very well, and we started talking about our families. He told me that his father was one of the founders of the St. John's School, a private school that placed a strong emphasis on academics and outdoor education.

I stopped in my tracks.

"Wait up, Mike. Your dad was one of the founders of St. John's? I read the book by James Raffan about that school just a few years ago!" I said. "They were nuts!" I caught myself. "I mean, ambitious."

"Whoa, Kyle," Mike looked at me. "You've read the book about St. John's? Barely anybody these days knows about St. John's."

"Ha! If there's a book in the library about canoeing, I've probably read it."

We laughed at the absurdity of him and me carrying a canoe in the Mantario Wilderness Zone, talking about his dad, the outdoorsman and the educator.

We grabbed my canoe and trekked back to Granite Lake. When we got back, we joined the rest of our crew for a swim. What a great morning.

We loaded up and headed off, everyone following my lead. I headed straight southeast to the bay at the far end of the lake, but when we got there we saw no white portage sign.

"Come on," I said. "We saw it yesterday. Where did the portage go?"

Patrick and Mike pulled up behind me.

"You led us astray, Kyle. The portage is on the western corner, not the southeastern corner."

"Why didn't you say anything?"

"We did. You just weren't listening."

Oops.

So we turned around and headed west. We turned the corner and saw the portage spot we swam at the day before.

Patrick and Mike were unloading their canoe while the rest of us floated close behind, waiting our turns. I peered into all the canoes and asked, "Hey. Is my fishing rod in any of your canoes?"

I heard four "No's." I looked in mine, and it wasn't there either.

"Arthur," I said. "When you were taking the load of paddles and life jackets this morning, was it there?"

"Yup. By the pile of paddles, right where you put it."

"Did you take it?"

"No. I thought you told me to leave it for you to get on the last trip."

"Arthur, I thought I told you to take it. Did you take it?"

"No, I did not Kyle."

I laughed. "You're pretty funny Arthur. Where is it?"

"Still there, as I said."

"Okay, last time. And no more jokes. I'm being serious now. Did you leave my fishing rod at Brant Lake?"

"Yes."

I put my paddle down and my face in my palms. I took a deep breath.

"Okay. We miscommunicated. I'm not mad. But I need to get that fishing rod back. Who wants to go back with me and get my fishing rod?"

Regan piped up from the front of Michael's canoe. "I will! Let's go!"

So we dumped our bags off at the portage into Sailing Lake, paddled back to the portage into Brant Lake, hiked 500 metres to the rocky clearing, removed wood ticks off our legs again, grabbed my fishing rod that was lying right where I had left it, walked 500 metres back to our canoes, and paddled back to the portage into Sailing Lake. I timed it and was impressed that it only took us twenty minutes. Regan was turning into quite the paddler.

When we got to the portage, all the boats and backpacks were gone except for ours. There was one cooler and a stack of paddles. We guessed that they had made a portaging plan, so we added our paddles to the pile.

"Regan, I really don't want to do this last portage in two trips. Want to do it in one?" I suggested.

"Yup. Let's do it."

We strapped our bags onto our backs, hoisted the canoe over our heads, and started walking. We ended up resting the canoe on our backpacks, transferring its weight to our hips, so it was a rather delightful portage. We even discussed Old Testament theology along the way! We passed a few others carrying their stuff, including Arthur and Connor standing by their canoe taking a break. But when we hit Sailing Lake, we were done with walking, and simply waited in our canoe for everyone to finish.

On Sailing Lake we paddled hard to the west, a north crosswind sending waves to batter the sides of our canoes. We gathered on the sheltered southern side of the island in Sailing Lake. After having some snacks there, we decided to reward ourselves with a flotilla, so we tied the canoes together and let the wind take us down to North Cross Lake.

We were moving at a good pace. Arthur and I acted as rudders from the back corners, while everybody else lounged around. Regan and Connor spotted the large green tarp folded up in one of the middle canoes, and started to conspire. They slowly stood up, stabilized by the flotilla, and each stepped on a bottom corner of the tarp while holding up a top corner.

The slack in the tarp disappeared, replaced by a wind-fueled curve. Our speed quickly tripled, and we marveled at how fast the shoreline whipped by. I stuck my paddle down in the water to steer and noticed the small wake our canoes were creating. This was the fastest I had I ever travelled in a canoe. We were all ecstatic, tossing snacks back and forth between the boats, grateful for Connor and Regan's willingness to be our masts.

Rounding a corner, our speed died with the wind. Alas, all good things must come to an end. We broke up the flotilla and put our paddles back in the water. Before us was the small dam between Sailing Lake and North Cross Lake.

After portaging for several kilometres the previous day, and starting this day off with a one-kilometre portage, carrying our canoes for 100 metres was like walking down the street to pick up our mail.

After a quick swim in the water flowing over the dam, and almost losing a pair of sunglasses and a bathing suit top, we were off. We camped on the west side of North Cross Lake, just past the dam, sheltered by the narrowing of the lake and an island. It was only mid-afternoon, and we were all glad to have had a shorter, fun-filled day. We set up our tents, went swimming, and ate supper four hours earlier than the night before. For those of us who didn't like portaging, this was camping at its best.

The real highlight for me, though, was Regan. If you recall, he was one of the "circle" paddlers from day one. He now asked me to go and teach him how to steer a canoe. My heart swooned.

We sat in my empty canoe in the water, me in the bow and Regan in the stern. We set the goal of collecting some firewood from a large beaver lodge in the next bay over, and after three strokes I knew he had figured it out. I could feel his J-stroke behind me, his anticipation of the canoe turning, and the gentle flicks of his blade to ensure our path would stay straight and true.

"Regan, you got this. You're paddling stern tomorrow."

His face lit up. Less so at my pronouncement and rather more so at his own sense of accomplishment, and how he had learned to paddle so well in just a few days.

We tucked in early that night, some of us happy to be going home the next day, and others of us sad the trip was ending. In hindsight, being stuck in the middle of two good things is a beautiful place to be.

Our paddle back the next day was picturesque. Clear water reflected the blue sky and puffy white clouds. The current in the tunnels was light, enabling an easy paddle through both of them. We stopped to swim on the island in South Cross Lake. By now everybody knew how to steer their canoes, so we made great time.

And, compared to the ten hours of paddling and portaging we did on the second day, nobody thought paddling four hours to get back to our cars was a lot of work. We passed day paddlers heading to the tunnel into North Cross Lake, and they admired us and the four days we spent in the Mantario Wilderness Zone. I beamed.

Near the end we paddled close together and I asked for people's highlights of the trip.

Regan jumped in first. "I learned how to J-stroke!"

"I'm glad we didn't meet any bears," said Mike.

"I learned that my body is strong," said Aly.

"I think I might love canoeing," mused Sam—which I already knew was true since she had asked to borrow my canoe the following day to come back to this exact spot with another friend.

After seeing almost nobody for the past two days, floating by so many cabins and boats on Caddy Lake felt strange. All the bright colours of the cars at the boat launch stuck out, as our eyes were used to seeing the greens, browns, greys, and blues of nature.

Running water in the bathrooms was a nice touch, though. As was the air conditioning in our cars. We strapped all the canoes to the trailer and buckled ourselves up. Some of us were quickly on our phones, texting our loved ones. Others took their time turning them back on, enjoying the freedom from their screens.

We had made plans to end our day with a late lunch at the Nite Hawk Cafe at West Hawk Lake. The ten of us filled a large table at the end of the restaurant. We delighted in the ice water and chocolate milkshakes served in massive metal cups, dripping with condensation.

We knew one of the servers there, but she wasn't scheduled to start work for a few hours. We texted her to come and join us. When she sat down, the first thing she asked was, "Did you hear about the bear attack in the Mantario Wilderness Zone?"

"What?" I said. "Black bears don't attack people."

"This one does," she said and proceeded to tell us how a hiker, only twenty kilometres from us, was bitten by a bear. Unprovoked, it had snuck up on him as he hiked the Mantario Hiking Trail, and bit him in the back of the knee. He was saved from significant injury only because he was wearing a hard plastic knee brace.

I listened intently, shocked because this bear was acting quite differently than most black bears, but also thinking about our night on Granite Lake, thankful that we sang Backstreet Boys as loud as we did.

After a 90 minute drive home, Aly, Sam, and I sent everyone on their way, thanking them for the great four days. We're quite confident that if we were to suggest another four-day excursion into the Mantario Wilderness Zone at least nine of the ten of us would sign up in a heartbeat.

Well, okay, maybe eight. Even I'll admit to moments on the portages that weren't all that fun. But at least we all learned that carrying a canoe over your head while wearing Crocs is a bad idea.

Acknowledgements

Thanks to Ashley Penner. Despite the many times that we've gone paddling together, early on in the process of writing this book I realized that you were barely in it. Sure, you hang around the edges of each story. But do you feature in one? Not really. Unless you count that time you rescued me by spotting us from the float plane. Or that time you threw a rope to Elisabeth and me to save us from the seaweed.

Upon further reflection, this could be because you are so competent and organized that, whenever we go paddling together, bad things don't happen to us. You remember to pack the maps, you don't take unnecessary risks, and you make sure that I start packing before 10:00 p.m. on the eve of a trip.

Sure, I guess you could frame this as "boring." But I'd rather frame it as "I love you, Kyle, and our kids, so much that I'm going to be the responsible one and make sure that nobody dies." This makes you the perfect paddling partner.

Thanks to Patti Enns. You drew all those maps for free, and STILL allowed yourself to be the punchline of the title. I will forever be grateful to you for bringing all these adventures to life. And don't worry. I promise to feed you if you become destitute.

Thanks to my parents, Allan and Simone Penner. We've never gone canoeing together, so you didn't make it into the book. But you did buy us our Wenonah Spirit II canoe as a wedding present. So without you, I'd have way fewer stories to tell.

Thanks to my editors and beta readers: Uncle Gilbert and Aunt Susan Brandt, Tamara Rempel, Cory Funk, Loretta Friesen, Sam Blatz, Paul Loewen, Garth Friesen, Allan Penner and Patti Enns. You helped edit my book without making me feel like a puddle on the floor. I am forever grateful.

And finally, thanks to all my canoeing companions: Ashley, Patti, David, James, Patrick, Phil C-E., Darcy, Kyle, Chris, Joanne, Kevin, Elisabeth, Werner, Mel, Becca, Sam, Sarah, Andrew, John, Phil P., Ryan, Paul, Arvid, Adrian, Julie, Caitlin, Jason, Warren, Kevin, Tom, Loretta, Paxton, Zander, Myelle, Tasha, Jeff, Nick, Alyssa, Mike, Michael, Naomi, Regan, Connor, Arthur, Arianna, Zach, and Milo. Thanks for going canoeing with me. I'd go with you again in a heartbeat.

Printed in Canada